Rate This Book

Rate This Book

How star ratings took over the world

Benji Wilson

new modern
ARCADE

new modern
ARCADE

First published in the UK in 2026 by New Modern Arcade
An imprint of Putman Publishing
Mermaid House, Puddle Dock, Blackfriars, London, EC4V 3DB

@newmodernbooks
@newmodern_books

Hardback ISBN: 978-1-917923-65-1
eBook ISBN: 978-1-917923-67-5
Audio ISBN: 978-1-917923-66-8

A CIP catalogue record for this book is available in the British Library.

Publishing and editorial: Pete Selby, James Lilford and Eleanor Porter
Typesetting: Marie Doherty

1 3 5 7 9 10 8 6 4 2

New Modern Arcade is an imprint of Putman Publishing
www.newmodernbooks.co.uk
www.putmanpublishing.co.uk

MIX
Paper | Supporting responsible forestry
FSC® C018072
FSC www.fsc.org

Printed and bound in Great Britain by Clays Ltd, Elcograf S.p.A.

To Caz, Stanley and Donnie.
Five stars. Best ever.

Contents

Introduction

So, how do you like it so far? You're eight words in. Fourteen now. And as nowadays nothing exists without a star rating to accompany it, it's time to start thinking about the end. How will you rate this book? Not *process* how – you know the process. When you've finished reading, or halfway through if you feel the need to, you can use the QR code on the back of the book, helpfully reprinted above, and rate this book in the usual way, from one to five stars. You can give it a gleaming endorsement on Amazon – your Kindle will ask you for it the minute you finish, using the exact words in the title, 'rate this book'. It only takes a second and then you can get back to whatever else you were doing – ideally, reading the rest of this book, or rating some other stuff that you've bought, consumed, visited or until this point completely forgotten about.

In the meantime, your appraisal will be aggregated into a mind-sump of easily comparable snap assessments. Maybe someone else, intrigued to discover why an author would be so dumb as to slap a crosshair on their work by calling it *Rate This Book*, will skim those ratings and decide to buy it. Maybe they won't (buy it, or even notice).

One thing is guaranteed: I will be watching, fixated on the starry average while spiralling down a plughole of insecurity and self-loathing. What I'm hoping to see is:

★★★★★

Look carefully at those humble, five-pronged stars and ponder: if you saw that in a review or on a poster what would you think? Would it vindicate your purchase decision, or indeed bolster your decision *not* to purchase and instead borrow your mum's Kindle? Would you secretly suspect that I have doctored the figures and pimped my own rating by paying my friends to vote hard in my favour? (I haven't – although if I drop below four, I wouldn't put it past me.)

Either way, stare at those five simple stars, at least for a second: what you're looking at is the Rosetta Stone of the information age, translating almost anything into a language of evaluation that we all understand. Because although people are well aware of the flaws of the five-star system – that it's unreliable, unmanageable and open to abuse – it has nonetheless become the de facto shorthand for appraisal across the world. This has taken fewer than 150 years, a piddling fragment of human discourse, and most of it has happened in a J-curve over the past 20. In a modern information society, where all of the details and data you could wish for are easily accessible from an electric pod in your pocket, when it comes to needing direction – what to choose, which brands and businesses to seek out – we have somehow reverted to the ways of the ancient mariners. We look to the stars.

Star ratings and reviews are the word of mouth for the modern era, but they are more than that. They are a compound of reputation and influence, both an illumination and a reflection, delivered with a tap and a click. They are mimetic and they are prejudicial; no one is immune to the pull-quote telling you the product or film or person has 25,000+ five-star reviews, even as you know full well that battery-farmed bots and commercial jiggery-pokery are almost guaranteed to be involved in such a chorus of approval.

Yet this system of summary, in as much as it *is* a system, has vanquished all of the others. Check your inbox. In mine right now, Sonos speakers are boasting about their panoply of five-star full houses. My children's school reports have just been sent to me and they're starred too. My Etsy order of a comedy golf club cover with a picture of my friend's face on it has been delivered and now they'd like me to rate the seller and leave a review. (I've given it a five. It's hilarious.)

User experience may vary, but it will undoubtedly involve five pointy things, because there is almost nothing left on the planet that isn't accorded a star rating: airport toilets, churches, art galleries, software updates, teachers, sneakers, posthumous Prince albums, Brazilian Bum-Bum Cream and, yes, you and me.

Part of my day job as a writer is giving out star ratings of my own. Mostly I review TV shows and, at my editor's command, I'm required to sort chunks of variegated television into one of five neat cubbyholes. I watch an episode, or a whole box set, sitting on my own with my headphones on. I distil pages of notes and hours of thoughts into a thousand or so words, and then I vacuum-pack all of those words into a standfirst (for instance, 'tedious pot-boiler best watched with both the sound *and* the picture off'). Then I distil *that* sentence summary into five identical symbols, or a fraction thereof. When you consider the months and years that go into making a TV show, the many people involved and the multiple viewers with vastly differing standpoints (some people *do* like Simon Cowell), it sometimes seems a little simplistic.

I asked my editor why we do star ratings at all. 'Because everyone else does,' she said. Although that line was delivered with an exasperated harrumph, it also points to a truth that undergirds this book: ratings are a self-fulfilling prophecy. They're as much about the rater as what they're rating. This is why star ratings and reviews have become ubiquitous in the age of data mining and online advertising: it's because what and how you rate tells someone else something about you. In this way, ratings accompany all modern commerce on both sides of every transaction. They give the consumer information about the product, but they also give sellers and advertisers information about the consumer. You'll have heard it before but it bears repeating: you are the product.

That's why the assessment boom has gone hand in glove with the growth of modern consumerism, as this book will show. Charting how star ratings took over the world offers a route map through the ways we choose to spend our time and our money, as well as the ways we allow our money to be taken from us. It also offers an insight into 21st-century credibility: who, what and why we trust in anything. Has the wisdom of crowds replaced the sureties of religion or academic expertise? Where does that leave what boomer sad-sacks with crappy Wi-Fi used to call 'truth'?

We will begin our spaceflight through the firmament by looking at how we got here. Chapter One shows how star ratings came about in the first place, why we use stars as symbols, why there are five of them and how they have become as much a part of modern discourse as emojis and hot takes. How has the Covid-19 pandemic and the corresponding decrease in human connection increased our reliance on a number between one and five for our understanding of the world? We rate everything and we get rated; a dropped star here or there can make or break products and people. Something that on the face of it is trivial has become essential.

Chapter Two looks at the psychology of ratings – what's really going on under the hood when you make a decision to do or buy something based on other people's appraisals, and likewise when you leave your own rating afterwards. What has this reliance on something reductive done to our ability to assess things for ourselves? Then, in Chapter Three, we move on to star ratings in the arts as arbiter and influencer, including controversies over what the stars mean and whether there's any better critic than the general public.

It's no coincidence that all of the big ratings aggregator sites use the word 'trust' on their homepage and Chapter Four looks at how the anonymity of the internet has made trust e-commerce's most valuable commodity. What replaces word of mouth when no one is actually talking to each other anymore? Why are we all putting our trust in faceless tech companies looking to maximise shareholder value instead of asking someone we know what they recommend?

Chapter Five looks at what happens when star ratings get personal. We've gone from rating products to rating people – doctors, lawyers, lovers – but they're not the same thing.

And then, if we're using star ratings to determine really important decisions such as which surgeon to take out our spleen or who to knock down our kitchen, then why aren't we using them for decisions like voting? Chapter Six suggests that maybe we should be.

In Chapter Seven I go back to where the idea for this book came from, which was when I discovered that I was being rated as a rater. Call it the Uber system – when ratings get reciprocal there's a whole new calculation to be made about what, when and how you're doling out judgement.

What if the whole game is rigged? Chapter Eight shows how the star-rating system has been weaponised and gamed. Reviews are now used as both bait and blackmail, and it is a major, worldwide problem.

Finally, we'll ask how if this system is broken, what might replace it? And, of course, I'll end up by fishing for a good rating myself again, in the process looking at how books like this one live or die by their own star rating.

What's incontestable is that, in the 21st century, star ratings and reviews – both reading and giving – have become the lingua franca of almost all human estimation. They are the lodestar for decision making which, given that executive function and rational choice are two of the defining features that mark us out as human beings, makes them very important.

Just ask Bethany Baptiste, Molly X. Chang, K.M. Enright, Thea Guanzon, Danielle L. Jensen, Akure Phénix, R.M. Virtues and Frances White. Their experience sends shivers down my spine as I type: all of them are authors who, in 2022, received scathing one-/no-star ratings on Goodreads for their exciting new books, and none of them knew why.

The lambastings turned out to be the work of Cait Corrain, a first-time author who had decided to use her literary chops to set up fake accounts and review-bomb her rivals. Her slights on her peers ranged, in her own words, 'from kind of mean to downright abusive' before she was outed.

The literary scandal (which was admittedly one of the best reads on Goodreads in years) led to Corrain posting an apology, but that didn't stop her being dropped by her agent and having her book deal cancelled. She told everyone not to read other books and, as a result, no one read hers.

And so, once more, with trepidation:

In asking you to rate this book, I am fully aware that I am tap-dancing in front of a firing squad. I have to contemplate the possibility that you just don't like my writing, my book, the cut of my jib or all three. I'm also aware that if in the past I've given you a kicking for a late-arriving Vinted parcel, or spurned, slighted or upset you in some other star-crossed field, then this might be payback time. A third option (also a common play in the starred society) is that you're shooting for the lols and my book is just collateral – just like the restaurant review that said, 'The banana bread pudding just tastes like bananas. 1 star.' I can already see it. Rate This Book: bantz67 writes, 'No stars. Bro asked for it.'

Whether you come to bury or praise, I hope I will have offered some understanding of why the world has quietly, possibly unthinkingly, given itself over to an abstract, unregulated scheme of measurement that is nonetheless now as universal as hours, minutes and seconds. I'm not sure I like the system, but like you, I'm caught within it: please do rate this book.

★★★★★

Chapter One

Stars Aligning – How We Got Here

Your opinion matters. It always has. The instinct to assess and grade, compare and quibble, is deep-seated and enduring. Ranking, rating, hailing and slating are skills that all lie hardwired into our evolutionary psychology. Classification is at the root of both knowledge and competition; saying what's what has always been closely followed by saying which is best. It is a longstanding strategy to help us make sense of a chaotic world.

As long as there has been art and culture, we have lauded some of it and taken immense pleasure in trashing the rest. In Ancient Greek theatre, art was always rated – trios of tragedies were put on in annual festivals and a winner chosen by juries, a bit like the Oscars.* The subject matter of these tragedies, with Sophocles, Aeschylus and Euripides their best-known authors, was very often democracy itself: for example, whether Creon, in Sophocles' *Antigone* (written around 440 BC), should rule 'for others, or for myself'. And the vote, that foundation of democracy, is the simplest rating system of them all.

---

* The plays, and later the actors, were rated by ten men, chosen at random from each of the administrative districts of Athens (and sworn to impartiality). The winning playwright (and actor) each earned themselves a place in the following year's festival.

The great Greek comedies took aim at Athenian aristocrats and encouraged the audience to laugh at their pomposity. Judgement – criticism – was not only allowed, it was part of the point. Though a group of judges had the final say as to who won, the audience was also encouraged to rate what they were seeing. They came up with a scratch system of applause, booing, hissing, belching or just falling asleep.[1]

Rating things, like any competitive sport, is fun to do, but it also appeals to another human instinct that's grounded in social psychology: hierarchy. We love to put things in boxes with labels, to codify and tabulate, to dither and squabble, and finally to determine which is better and who or what is best. Sociobiologists will tell you that humans appreciate hierarchies because once one has been established, we can place ourselves within it. Hierarchy can be hell if you find yourself dumped on the lowest rung of the ladder, but in evolutionary terms, it has at least offered us flailing Homo sapiens some structure. Species like humans gain security and safety by knowing their place.

In *History of Animals*, written around 350 BCE, Aristotle, Athens' original busy-bee polymath, turns his hand to zoology, attempting in 300 pages nothing less than a comprehensive classification of every living thing. It's a daunting undertaking, one full of brilliant insights as well as (in hindsight) some dur-brained howlers. But underpinning it all is an inherent snobbery. Aristotle approaches the world with qualitative specs on: the unwritten assumption in his *History of Animals* is that some animals are just better than others. (He prefers 'gregarious' creatures, for example, as if life on earth is just one big dinner party.)*

---

* By 'gregarious', Aristotle meant living in flocks or herds, as opposed to being solitary. But he distinguished between gregarious and social behaviour, with gregariousness being apolitical and not working towards any common goal. My ancient Greek is not what it was, but I like to think that Aristotle still liked animals that were kinda chatty.

The Aristotelian grade card ranks and reviews all things in existence from one to 11,* with plants at the bottom and the eleventh rank ('perfect') reserved for life that is warm, wet and quite similar to its mature form at birth. He therefore reckoned that humans were the very best of all things (and he hated eggs).

Aristotle wasn't writing to help bemused consumers pick their favourite animal on Trustpilot, or to settle some internet tiff. To him, ranking some things as better than others (with humans at the top of the heap) was just an expression of what he called 'the great chain of being'. Implicit in that chain was that it ran vertically – it had a top and a bottom. The world as we find it, Aristotle thought, contains an innate classification that we can and should demarcate and try to agree on.

Aristotle's systematics were based on proliferation: he saw a lot of animal life all around him and he figured it needed categorising. In fact, feeling overwhelmed by what we now call information overload and wanting to establish some cognitive clarity runs right through Plato (who, via Socrates in *Phaedrus*, thought that even writing things down would weaken memory and lead to an illusion of wisdom rather than true understanding), Seneca the Younger (the Roman stoic philosopher who called the abundance of books in the world 'a distraction') and on to TS Eliot ('Where is the wisdom we have lost in knowledge/ Where is the knowledge we have lost in information?' he wrote in his pageant play *The Rock* in 1934).

But the path to a society in which almost everything is bracketed, reviewed and ranked – the society where this book is set – requires the luxury of choice, as well as a large body of people all looking for

---

* Aristotle liked elevens. His prescription for a full and flourishing life across the *Nicomachean Ethics*, *Rhetoric* and *Economics* consists of eleven 'Excellences', including Excellence in Judgement. Had he indulged in star ratings, he would probably have plumped for a Nigel Tufnell-esque scale of one to 11.

direction. Both of these – a multitude of things to choose between and a multitude of bewildered choosers – in turn required new technologies. To take an apposite example, you didn't need a five-star book review until there were books ... and then lots of books ... and then lots of people reading them. Spooling back through history, all of that leads back to Gutenberg's invention of the printing press around 1440.

Essentially, human beings had to have choice to need guidance and, for most people from early history via medieval times and on to the dawn of mass production, choice was an unimaginable indulgence. One chamber pot was most likely better than another, or some brooch-makers or blacksmiths superior to your local one. But you couldn't travel to try them out and a/b comparisons only apply when you have a 'b' to compare with your 'a'. And, for many hundreds of years until the industrial revolution and the age of mechanical production, most people didn't.

The other necessity for ratings and reviews is a means of communication to get them out there. Here again, none existed until the birth of the mass media, with the first newspapers in English appearing in the 17th century. Before that, star ratings would have to have been carved on a rock.

What about the theatre, that hotbed of early criticism and voting with your feet (and rotten fruit) as mentioned above? It didn't survive the decline of Athens and then Rome, and was virtually wiped out as the Church increased in power and influence. In the Middle Ages, plays and stagecraft did re-emerge as a means to tell Bible stories (culminating in the annual nativity play that, speaking as a former Sheep Number Three, still haunts school halls every year). But who, faced with the power of organised religion, was going to say that a mystery play or the nativity was a dreary old dirge, one star, don't bother?

Elsewhere in the arts, however, evaluation and analysis continued to go hand in hand. The Villani brothers' *Cronica*, written across

56 years from 1308 onwards and often seen as the birthplace of the critical tradition, established (among many, many other things; it runs to 12 volumes) that painting was a superior practice to other liberal arts, and that Florence, where Filippo Villani happened to live, was the acme of all civilisation.

In the 15th century, Ghiberti's *I Commentarii* described art as a practice that was an irresistible frogmarch of progress, not just changing or evolving but getting better and better. Some art is just more equal than others, Ghiberti insisted, and early art criticism asked you to trust some critics more than others too.

The most famous work of the period, Giorgio Vasari's 1568 *Lives of the Most Eminent Italian Architects, Painters, and Sculptors*, took things a step further, offering its own very Aristotelian criteria of value before using the author's own blueprint rating system to divvy up Italian renaissance art.

Vasari, who lived from 1511 to 1574, looked at the three preeminent art forms of his time – architecture, sculpture and painting – giving readers the biographies of artists from the Middle Ages to his own day, with a focus on their creations. In the Preface to Part Three, he takes a step back from his history in order that he may 'define briefly' the qualities that he thinks make for good art: 'good rule, order, proportion, design, and style'.

Pleasingly for this book, there are five grades, with Vasari using his newly minted criteria to assess individual artists as well as entire periods. Thus, the 14th century is, to Vasari, a bit of a dead zone, the 15th century is slightly better, but the 16th century – his own time, as it happens – is the bee's knees. He labels those 300 years the bronze, silver and golden ages of art respectively.*

---

* Vasari didn't invent this now much-used (and over-used) grading. The term 'Golden Age' originates from Greek mythology, first described by

Crucially, Vasari was himself a painter and an architect (and, like Villani, he lived in Tuscany, the wellspring of the Italian renaissance). Whatever readers made of his judgements, there was reason enough to trust them. That trust is another key component of the who-rates-the-raters nexus that we will later revisit.*

With a grammar of criticism established – some things are better than others; they can and should be categorised according to a series of ascending plinths; so-and-so is the right and true person to do that classification – it nonetheless took several centuries to get from Vasari lionising Michelangelo to the 4.9 star rating that Michelangelo's David sculpture currently gets on Google Reviews.†

The Enlightenment search for order – which reached its apogee in Linnaeus's *Systema Naturae* (1735), Samuel Johnson's *Dictionary* (1755) and Diderot and D'Alembert's whopping *Encyclopédie* (published between 1751 and 1772 in 17 volumes) – set the path towards a reviewed, ranked and rated society.[2] D'Alembert, in his preface to the *Encyclopédie*, spoke of the complexity of the task he and Diderot had set themselves.

'The general system of the sciences and arts is a sort of labyrinth, a tortuous road which the intellect enters without quite knowing what direction to take,'[3] he said, instinctively reaching for the language of the lost traveller in search of a guiding star.

Hesiod in *Works and Days* (c. 7th century BC) as the first and best of five – of course – descending, chronological ages of humanity. Below bronze were the frankly godawful iron and stone ages.

* Vasari's status as one of the first ratings influencers is shown in the VASARI ratings system used by radiographers to assess tumours in MRI scans. It stands for (Visually AcceSAble Rembrandt Images), but shows that when it comes to criteria, Vasari was first.

† David doesn't get a five thanks to reviews like 'local guide' JST's from mid 2024: 'Overrated. The word is CHAOS. Just to see a marble man with tiny balls.'

Linnaeus's system was rank-based. He somehow managed to get through more than 12,000 species, but his classification was ruthlessly hierarchical. Once again, Homo sapiens were at the top of the pile.

A few Enlightenment eggheads on the search for knowledge, however, didn't require a universal ranking system to disseminate their findings. Star ratings and reviews as we know them today needed three things to flourish: a profusion of stuff to be rated, someone to rate them and a glut of time-poor readers who wanted quick summaries. It therefore follows that the modern ratings game only really began in the mid-to-late 19th century, as a sucker shoot from the flowering of the consumer society. It took the rise of the middle-classes and the things the middle-classes like to do – buy stuff, travel for fun and read newspapers and books, but at the same time be really very busy all the time – for Aristotle and Vasari's yes-but-which-is-*best* instincts to find their fullest expression.

★

The roots of compulsive starring lie in that great middle-class obsession: travel guidebooks. These came into being at the same time as the whole concept of consumers and consumer choice.* As the middle-classes and the aristocratic leisure class merged, all financed by the spoils of the industrial revolution, and as travel went from being an occasional necessity to a breezy past-time, two things changed.

First, sites of interest, and then hotels and restaurants, became something to be weighed up and plumped for. Your dandy traveller had options. Then, if there were choices to be made, those options needed to be graded. Once the concept of choice and rank filtered down to the

---

* The word 'consumer' meaning anything more than someone who eats or uses things only came to prominence at the beginning of the 20th century, according to Google's Ngram concordance of published books.

mass market, each and every potential consumer decision was ripe for grading too. Ratings offered a beacon line through the rapid increase of products and services that followed the industrial revolution round Europe like so much flotsam.

But how to grade? Although Vasari's gold, silver and bronze had been used since the Ancient Greek poet Hesiod as decorations in accordance with their actual material value (see footnote above), in print they were problematic. Three tiers was, as we shall see, not enough to make sense of a morass of options. If concision was the goal in this new age of paper and ink, writing out the words 'gold', 'silver' and 'bronze' again and again was hardly a space saver. There was no way of representing them by colour, as there was no colour printing. Something that fell easily to hand on the newly invented typewriter (1868), or a glyph that was a staple of the printer's block,* was required.

A solution began to emerge and, like so many eminently sensible ones, it was created by women. The Home Tour was a series of routes and sites in Britain in the 17th and 18th century, labelled in contrast to the flashier Grand Tour that people in Tobias Smollett novels liked to partake of around Europe. Women were much more likely to undertake the Home Tour – which might take in the Peak District, the Lakes, Scotland or North Wales – than the swanky foreign one; it was cheaper, quicker and more practical. As such, the Home Tour produced a number of manuscripts and published accounts detailing its writers' experiences.

It was little-known Home Tour writers like Elizabeth Diggle, Caroline Lybbe Powys and Dorothy Richardson who first came up

---

* In 1885, the American inventor Tolbert Lanston applied for a patent on a typesetting system that included the basic Monotype keyboard. This included a star or asterisk symbol. See Claire Cock-Starkey's *Hyphens & Hashtags*: *The Stories Behind the Symbols on Our Keyboard.*

with the kind of structural mechanisms in their handwritten accounts – mileage, maps, itineraries, prices – that we now recognise as the fuselage of modern guidebooks.[4]

Symbolical rating systems were one of those structural mechanisms: Elizabeth Diggle's 1788 manuscript travelogue *Journal of a Tour from London to the Highlands of Scotland*[5] consists of a series of letters to her sister which were transcribed and bound together in the form of a travel journal. Towards the end, Diggle turns Tripadvisor, rating the inns she stays at from 'very bad', 'tolerable' and 'indifferent' to 'very good' and 'excellent'. Hers is the first five-tiered ranking system. Scottish inns of the period come out of it surprisingly well.

It took a writer called Mariana Starke to take the methodology of the Home Tour women and get it into print, in the process showing the mass market how this new shorthand might work for them.

Born in 1762, Starke was a lady of letters, writing plays and poetry as well as translating, but she also had an eye for a business opportunity. Aged 30, her whole family had decamped to Italy overland via France, looking for the healing airs after every single one of them bar Mariana contracted tuberculosis. As with Keats, the wondrous continental climate failed to save her younger sister or her father, but it did give Mariana an idea.

On her travels, she had noted that she was one of many peripatetics. A work that 'comprehends every kind of information most needful to continental travellers' might, she thought, support the 'immediate & immense' emigration that Starke saw would follow the end of the Napoleonic Wars.*

---

* She was right. It was the return of peace in 1815 that triggered a pan-European migration wave, as the economic safety net of wartime vanished and returning soldiers struggled to find work at home.

'Allow me to enquire,' she wrote to the publisher John Murray, 'whether, if that event occurs, you would like to purchase the above-named work?'

Starke's move into travel writing was therefore based on a punt: that if only England and France could stop their bickering (quite a big if, considering the previous 500 years' history), a new breed of tourist would want to sample the delights of continental Europe. In order to be 'reliably delighted', she reasoned they would need a reliable (and delightful) guide. Mariana, who called herself 'a faithful historian', reckoned she could be that guide: the original trust pilot.

Starke's *Travels on the Continent* (1820) and *Travels in Europe* (1828), both published by Murray, are some of the first popular guidebooks, spawning a genre that now takes up entire bookshops as well as countless petabytes of server space. Her great insight was that 'directions' (a word she added to the title of later editions of *Travels on the Continent*) was a term with several meanings. In the 1820s, a new breed of moneyed consumer needed not just *directions* how to get somewhere, but *direction* as to whether it was worth going in the first place. Her books were written not merely for edification or enlightenment, but as pocket-sized tools. They were meant to be useful.

Almost immediately, Starke found that, as a no-filter kind of observer, she had an abundance of material. Too much, in fact. For instance, she loved art and there were vast amounts of things to see in France and Italy, but Starke recognised that her readers would be overwhelmed by a bow wave of relentless enthusiasm. Just like the compass points used for geographical directions, she needed a shorthand for her form of 'direction'. She needed a rating system.

'As considerable changes have lately taken place in large continental cities relevant to large works of art, I have endeavoured, in the following pages, to furnish Travellers with correct lists of the objects best worth notice in the respective galleries of sculpture and painting; at the same time marking, with one or more exclamation points (according to their merit) those works which are deemed particularly excellent.'*

Her exclamation marks, she insisted, were 'for the benefit of persons restricted in point of time'. The typography itself – a sort of Edwardian Face-with-Open-Mouth emoji that screamed 'LOOK!' – allowed you to skirt across the page and cut to the chase.

The peregrinations that follow quickly start to look like a teen-ager's diary, with exclamation marks spread across the pages like squashed flies on curtains. By page 13, Starke has already hit a six-exclamation-mark double-top for the 'transcendent and inimitable' *Borghese Gladiator* sculpture in the Louvre.†

But just as she invented the ratings system and established that six was the pinnacle – arbitrarily; she never says why – Starke quickly becomes aware of its pitfalls. As the book rambles through Italian antiquities and masterpieces, she reins in her vibrant exclaiming. Nothing else gets a sixer, even as she gets to the Uffizi, the Vatican and Vasari's matchless Michelangelo. In fact, after an initial orgy of ratings, she ends up canning it altogether, perhaps aware that machine-gunning ratings willy-nilly dulls the effect. Both of these – don't shoot too high too soon; don't shoot too much – remain star-rating red flags today. Starke just found them first.

The origins of star ratings and reviews therefore lay in a specific need: people had the means and the desire to travel, and there was a profusion of choice as to what they might do, so up sprung Mariana Starke with a salve of typographical ease.

---

* Anyone who has ever lugged around a well-thumbed Lonely Planet guide will recognise this system.

† The *Borghese Gladiator* has in fact proved very imitable. Everyone from Giordano to John Singleton Copley has produced works copying the pose.

It's worth noting that Starke not only came up with the denominational shorthand, but she set a tone as well: she is an enthusiast, a praiser, not a knocker. In *Travels*, one exclamation mark is still considerably better than no mention at all. It's just not as good as two. But that's also pretty darned good.

Most noteworthy is that Starke recognises that, for any of this to matter, she needs to earn her readers' trust. She is aware that families will be travelling on budgets and recommends accommodation 'for persons who understand and practise economy'. She knows that travel can be dangerous too, but assures her subscribers that, thanks to the recent introduction of street lamps, 'the dreadful practice of assassination' has been stopped. In short, in order to be trusted and give out her own ratings, she first requires readers to trust her, just like Vasari.

But asking readers to trust you is also putting your head above the parapet. From the get-go, rat*ers* have quickly found themselves rat*ed*. Starke's success with her exclamation marks led to a summary slapdown from the author of the satiric 'Verses Written in a Blank Leaf of Mrs. Starke's "Guide for Travellers"', published anonymously in the periodical *John Bull* in 1836:

'She calls Italians "civil" / Yet says, "they often cheat;" / And that you "first must bargain" / For all you buy and eat'.

The criticism seems to be that Starke's assessments are inconsistent – another common barb among star rating and review systems today.

A further danger of offering guidance is that some interested parties won't like what you're saying. In a letter to her publisher, Starke noted that her suggestion that some tour guides in Rome were ripping off tourists hadn't gone down well: 'Another person wrote me a letter protesting that if I persevered in opening the eyes of travellers with respect to expenses on the Continent I might expect to be assassinated.'[6]

Influencers and opinion formers will recognise this tossed-off death threat as the dawn of below-the-line commentary. Mariana

Starke wanted to show what was 'worth notice' and she thought readers would notice a bunch of exclamation marks sitting in the middle of the page. She was right, but in so doing, she got herself noticed too.

The road to starry ubiquity followed the newly emboldened, newly quite wealthy traveller round Europe. Starke's prototype travel guide was a hit, running to several editions and being read and copied both by her own publisher John Murray, who released a series of hugely popular Handbooks for Travellers, and then by the German Karl Baedeker. In 1846, in the third edition of his *Handbuch für Reisende durch Deutschland und den Oesterreichischen Kaiserstaat,** Baedeker took Starke's exclamation marks and swapped them out for stars, in the process expanding their purview to sights, attractions and accommodation.

At first, Baedeker merely offered a single star printed alongside the names of hotels that floated his boat. But, in 1844, he expanded his range, taking in visitor attractions more generally and upping his top grade to three stars. It was Baedeker who reasoned that if travellers wanted help assessing whether the Sistine Chapel was better or worse than the Louvre, then they might also appreciate a ruling on the spittoons at the George V or the plush, snowy carpets at the Savoy.

Baedeker's worldwide success made his surname synonymous with travel guides, but it also introduced the world to stars as a common code.† When the British playwright George Bernard Shaw wanted to

---

* *Handbook for Travellers Through Germany and the Austrian Empire.*

† One unfortunate corollary was the so-called 'Baedeker Raids'. In spring 1942, the German Air Force launched a series of devastating strikes against several historic cities in Britain. Germany's decision to attack British cultural centres was a response to the bombing of its own historic towns by the Royal Air Force. Nazi propagandist Baron Gustav Braun von Stumm said: 'We shall go out and bomb every building in Britain marked with three stars in the Baedeker Guide.'

have two characters discussing the merits of a local landmark in his 1892 play *Widowers' House*, he has one of them ask, 'Is the church a celebrated one?' and another answer, 'Baedeker stars it.' Starring something had become a recognisable verb of commendation. The *OED* dates this usage of 'star' to the mid-19th century with the given meaning of 'Distinguish by an affixed asterisk; single out for special notice or recommendation.'*

The star or asterisk had replaced the exclamation mark or gold medal as the unit of complimentary currency and we might pause here to ask why. There is no record as to how Baedeker alighted on the star as opposed to any other symbol, though presumably he thought it best to change from Starke's symbol or be accused of copying her (which was precisely what he had done).

In Baedeker's defence, asterisks (from the Greek 'asteriskos', meaning 'small star') have been used since medieval times in text and printing as a distinguishing or dividing mark. Often they pointed to an omission or a footnote,† implying 'there's more to be discovered here.'

Anyone can draw an asterisk. They're symmetrical and celestial and, more importantly, easily available – the asterisk was ready for deployment in some of the earliest printer's cases, generally at the very top left of the 'upper case'‡ as an obvious first pick. The asterisk has been ever-present, first in the standard Letraset typeset and then in the Monotype matrix. It sits proudly above the number 8 on a US standard QWERTY keyboard, it was there in the first ASCII American

* It's worth nothing here that the *OED* also lists a definition of the word 'rate' as 'Chide, scold, reprove angrily (a person or a dog)' from late Middle English. 'Rating' has always come with an implicit threat of disapproval.

† Like this one.

‡ The phrases 'upper case' and 'lower case' derive from the two sides of a printer's case where the letters were stored.

Standard Code for Information Interchange (the original character encoding standard for electronic communication) and it was part of the Unicode Standard* that followed it in the early 1990s.

And the symbolism works. Stars stand out, offering guidance amidst the gloom. They glitter and shine in the dark night. Good things are 'written' in them, whereas an ill-starred person is one destined to fail. What could be more apt as an emblem of distinction?

★★

Once Baedeker had established stars as a universal commendation to go with his reviews, it was motor cars that drove them around the world. In 1900, the French tyre manufacturer Michelin realised that, in order to sell more tyres, the company needed to get more people into the then new-fangled apparition known as cars. Many motorists – another consumer class born of a new technology – were pootling around in their magnificent automobiles, but then getting hopelessly lost or having trouble finding reasonable meals and accommodation. As such, brothers Édouard and André Michelin thought they would publish a giveaway pamphlet full of Baedeker-style hints and tips in order to get people wearing out their tyres – and thus needing to buy new ones.

Over in the UK, the Automobile Association had done something similar in 1908 when they had also begun assessing hotels and restaurants. Four years later, AA secretary Ernest Stenson-Cooke decided to borrow the star system used for rating brandy in order to classify hotels.†

* A text-encoding standard maintained by the Unicode Consortium and designed to support the use of text in all of the world's writing systems that can be digitised.

† In brandy, stars merely signify years stored in cask. A three-star brandy has been stored for at least two years; a five-star for four. Perhaps it makes sense if you've drunk a lot of brandy.

At least that's what he said, though it's hard to believe he'd never seen a Baedeker at that point.*

By 1926, the Michelin brothers had followed Murray, Baedeker and the AA in introducing stars for restaurants; the French were very picky about what and where they ate. Initially, this was a binary system where a star meant 'go there' and no star meant 'don't bother'. Five years later, in 1931, a further classification arrived by customer request. One star was still good, but two represented 'merits a detour' while three 'vaux [justifies] le voyage'. In every case, the question was whether the meal was worth the trip.[7]

Two things emerge from the early history of what is now the most famous star rating system in the world. Actually, three. Firstly, Michelin is a three-star system, not a five; secondly, that three-star rating is for the most part withheld and therefore more valued; and, thirdly, my oh my there's money to be made in stars. In the 1954 Michelin guide, there were only 12 three-star restaurants in all of France, prompting the *New Yorker* to note that 'in France, a man with a three-star restaurant grows rich, a two-star man will prosper and the man who loses his single star may go out of business'. It was the accumulation of Michelin stars that made the Roux brothers, and then Gordon Ramsay, Thomas Keller and Heston Blumenthal very wealthy. Stars glisten for a reason – there's gold in them.

In 1922, just before they introduced stars, the Michelin brothers started charging for their guide (around £1.60). 'Man only truly respects what he pays for,' said Édouard at the time, implying that insight had become valuable to the point of actual monetary worth. People would pay for Michelin's reputation as an authority. That

* In 1955, the AA added a rosette system 'a subjective system reflecting the emergence of British hotel cooking after 15 years of rationing.' A three-tier system was used. In 1968, they added a 'five-pennant system' for caravan sites.

authority came hand in hand with every Michelin inspector's anonymity. Inspectors (the title implies some Poirot-style fine-tooth-combing to try to catch restaurants out) came with the cloak-and-dagger frisson of a Catholic inquisition. The *Guide Michelin* gained its popularity and much of its considerable heft from its painstaking, unsentimental ranking of places to eat.

'The *Guide* has grown so much in recent years that it now has practically dictatorial status,' said a mildly overawed M. Thuilier in the *New Yorker* in 1954. 'Its prestige is such that even a man who carries a copy commands respect.'

By the middle of the last century, therefore, Michelin's stars had changed both rating and getting rated. Six years before the very first bubblings of the internet (starting in 1961, US government researchers began attempts to use room-sized, seven-tonne computers to share information), the Michelin guide was already a bellwether as much as a map.

What's more, merely purchasing and carrying a red-covered *Michelin* upped the purchaser's own societal rating. It implied sophistication, street smarts, class. As for the hotel trade itself, Michelin had – on its own three-starred terms, without external regulation or accountability – become an unsparing authority. 'The *Guide* takes no back talk,' wrote Thuilier, 'and is not impressed by reputations or influence.'

This ratings game was a serious business, but if you were one of the winners, you could use your stars as promotional grist. Other guides wanted a piece of Michelin's tarte, one-upping the French guide's USP by doling out not merely three but four and latterly five stars.

In the US, for example, the American Automobile Association had been running field inspections of restaurants and accommodation since 1937 as a service to their members. In 1963, it adopted a formal rating system for its Tour Book guide accommodation listings.

The 1963–1964 editions of the *Florida* and *Southeastern Tour Book* guides were the first to carry these ratings, which included four specific categories: Good, Very Good, Excellent and Outstanding.*

In 1976, the early lodging rating system evolved into the diamond rating system, which has five categories (if you include the category of not selected as the equivalent of no stars). They chose a diamond as AAA's rating symbol as a nod to the association's 75th anniversary (or that's what the AAA says; presumably the fact that Michelin and the UK AA had already plumped for stars was a factor too). In 1985, the AAA began handing out diamond ratings for restaurants.

As the AAA's 62 million members currently book more than 31 per cent of all paid room nights in North America, that's a lot of people who've grown up surrounded by a five-diamond system. Add to that the growth of the *Mobil Travel Guide*, an attempt by publisher Simon & Schuster to create an American *Guide Michelin* that ended up as the *Forbes Travel Guide*, and you have an available population of many millions of upwardly mobile consumers all schooled in the art of five-category, shimmering guidance.[†,8]

Other countries have fallen in line – although never quite the same line. Since 1999, 19 European nations, including Germany, Greece, Denmark and Switzerland, have signed up to the Hotel Stars scheme,[9]

---

* And, of course, it's not actually four rankings, it's five. Never forget the silent slur of not being rated at all.

† The *Forbes Travel Guide* has a suitably meandering history. The authors of *The Stevens America: A Traveler's Guide to the United States* (1950), Marion and Alden Stevens, asked Max L. Schuster of publishers Simon & Schuster if they wanted another travel book. Schuster was a big fan of the Michelin guides and so offered the Stevenses a chance to create a US version. Mobil Oil agreed to pay for the project, and so in 1958 the first *Mobil Travel Guide* was born. In 2009 it was renamed the *Forbes Travel Guide*. Print publication ceased in 2011 and the *Forbes Travel Guide* is now entirely online, like everything else.

another one-to-five system designed to standardise hotel classification. Inevitably, other countries then stuffed up that attempt at streamlining by introducing their own classifications, such as the seven-star Burj Al Arab in Dubai and the eight-star Emirates Palace Mandarin Oriental, Abu Dhabi, which opened in 2005.* As the number of stars has headed skywards, at least in hotel ratings, the standard classifications have become correspondingly meaningless. But the supremacy of the star system is why schemes like hotelstars.eu keep trying.

You can see how pithy, authoritative and reliable guidance for the weary traveller and the disoriented motorist might prove useful in other fields. If all life is a journey (a trope that dates back to Middle English, but that more recently has been hammered into incomprehensibility by competitive talent shows), then everyone needs some roadside signage. As the doctrine of people as 'consumers' – rational economic actors making decisions aimed at maximising their personal utility or benefit – spread throughout the 20th century, so the ratings and reviews required to make those decisions with alacrity and purpose spread too.

★★★

The agony of choice was soothed by the glistening balm of the stars. In the mass media, this began in books, the first mass market consumable with a long-term critical tradition. Edward J. O'Brien's *The Best American Short Stories*, published annually from 1915, used a rating system of zero to three stars based on 'literary prominence'.[10]

* When Dubai's Burj Al Arab opened back in 1999, it was dubbed the world's first seven-star hotel by a journalist overwhelmed by its unparalleled opulence. The moniker stuck, despite the hotel claiming it never wanted to be called seven-star.

O'Brien's starting point was adversarial. He was fed up with critics, mostly English, saying that American literature was crap. His introduction is a call to arms:

'We have listened to much wailing during the past year about the absence of all literary qualities in our fiction. We have been judged by Englishmen and Irishmen who do not know our work and by Americans who do know it. The general verdict against American literature is Guilty! Is this wise? Is this just?'

No, it was not just, not to Edward J. O'Brien. The whole purpose of his appraisal was to stick one up those toffee-nosed Brits, but in order to do so, he felt he needed to tool up. In his introduction, he recognises that the problem was one of volume: 'The main trouble with the fellows who are writing in America to-day is that they write too much.'

What could he marshal as both sword and filter? He chose to wade in with three asterisks (or more accurately four, as stories 'which fail, in my opinion, to survive either the test of substance or the test of form' could be empty-chaired and given no asterisks at all). The scale then ran all the way up to three-star doozies which possess 'the distinction of uniting genuine substance and artistic form in a closely woven pattern with a spiritual sincerity so earnest, and a creative belief so strong, that each of these stories may fairly claim, in my opinion, a position of some permanence in our literature as a criticism of life.'

Note the multiple uses of 'in my opinion': O'Brien tries to outflank future critics by admitting that when he ascribes an asterisk or more, 'this asterisk, I must confess, reveals in some measure a personal preference'.

This is hardly a confession – of course rating and reviewing something reveals a personal preference. What he perhaps didn't see coming was that people *liked* personal preferences, liked having complex things boiled down to simple numbers, they liked using them as a punchbag for disagreements and a launchpad for their own personal preferences.

The Anglo-American writer Oliver Herford's personal preference, in his travel book *Neither Here Nor There*, published in 1922, was to take aim at Edward J. O'Brien and his asterisk ratings. Herford hated them.

'Never, I think, were a mob of overworked employees so pitifully huddled together in an ill-ventilated factory as are the Asterisks in this Sweatshop of Twaddle.'*

Yet whether or not asterisks and stars are malnourished children working in a sweatshop of twaddle (which, by the way, is an excellent name for anyone thinking of a new aggregator site), their twaddlishness proved instantly popular. Anywhere that discernment crossed paths with consumer choice soon found itself part of the panoply, as stars alongside reviews seeped from arts criticism in travel books into restaurant and hotel guides and eventually back into the popular arts again.

Cinema, that modern cathedral, was next. In 1928, star ratings self-seeded from the written word to the moving image, when films' merits were first quantified in the *New York Daily News*.

Three stars meant 'excellent', two 'good' and one star meant 'mediocre'. Yet again, there was a fourth category – no stars – though this time it was unambiguous (the paper's critic Irene Thirer said that no stars 'means the picture's right bad').

In 1928, it was only 33 years since the Lumière brothers had first projected moving images to a paying audience in Paris, a year since

* Not only did Herford have a pop at O'Brien, but he also lashed out at *Baedeker's Guide to Berlin* for the crime of disagreeing with his own artistic sensibilities: 'A well-preserved Archaeopteryx in Solnhofen state, to me by all odds the most interesting object in Berlin, has no star at all!'

The Jazz Singer had brought in synchronised dialogue and four years before Technicolor. But the notion of a fixed scale of appreciation came just in time for cinema's first golden age when movies became the principal form of popular entertainment. Needless to say, the first film review in the *New York Daily News* was one-starred (*The Port of Missing Girls* being the object of blithe dismissal).

Not that the star system took off straight away in movieland. It didn't appear again until the 1950s, back in Michelin country when the newly founded French film magazine *Cahiers du Cinéma* started polling critics and indulging in a spot of what we now call aggregating, Rotten Tomatoes-style. In the finest French tradition, *Cahiers du Cinéma* would print a bullet next to movies that the magazine particularly despised, such as *Lawrence of Arabia*.

Ratings were being weaponised. Stars shone, but they had sharp points too. They could snark as much as coddle, promote or demote, depending on their deployment.

And from the outset, the critics assigning these stars had issues with them. One side said they were a cumbersome, simplistic tool, like handing a pin-eyed sharp-shooter a blunderbuss. If you tagged a rating to your analysis, the argument went, then readers would just skip all that purple prose and jump straight to the label.

The other side pointed to the reader, aka the consumer. Lives were busy, and better to have a reader dip into your publication for that crucial yay or nay than not buy it at all.

'They do get you on to the poster, but I don't like them,' sniffed Susannah Clapp, a long-time theatre critic for the *Observer* newspaper in the UK, many years later.* 'They mean people don't read: they count.'

---

* The *Observer* was the last paper in the UK to introduce star ratings and still only gave them online, not in print. Formidable – though possibly

'It's so arbitrary as well,' chimed in Clapp's colleague and film critic Simran Hans, 'I've written so many three-star reviews that are quite negative. Or two-star reviews that contain praise. Most critics would rather not do them.'[11]

A fair point, and yet most critics *have* ended up doing them – because out in the real world, the birth of the blockbuster, the sequel and the advent of television meant that the consumer's life was just getting busier, their leisure-time choices more manifold (and hence less leisurely). All of this depended on these art forms being commodified. Star ratings looked good on posters (as long as you had four or above) and were an excellent cipher for advertising; if your mag didn't use them, then you weren't getting that free publicity on all those movie ad campaigns. Ratings culture spread like wildfire. Every magazine and newspaper had to take a position.

The theatre took some time to catch up, in part because theatre criticism was an older art (see Aristotle, above) and in part because the domination of screens meant that fewer and fewer people went to the theatre as the 20th century wore on. It also had an air as a more sophisticated art form, one that cocked a snook at snap judgements and the demands of the hoi polloi with their so-called 'busy' lives.

But, as *The Oxford Companion to English Literature* notes, 'the distinguishing feature of English-language theatre criticism remains a suspicion of intellectual theory and a trust in subjective impressions'. In other words, theatre criticism was ripe for getting starred-up because it was founded within an audience of belching or applauding individuals all furiously rating and reviewing what they had seen.

---

misguided – hauteur. In 2024, it was sold to Tortoise Media and now gives no star ratings at all ... in March 2026 it offered its entire staff voluntary redundancy, according to *The Times*.

Once it got there, theatre went to town on star ratings. By the early 1990s, the Edinburgh Fringe, the largest arts festival in the world, was serving up in excess of 1,600 shows a year for review and neither the punters nor the local newspaper, the *Scotsman*, could keep up. Star ratings were introduced in 1996 and the UK's first theatre website, Whatsonstage.com, duly added stars to its reviews, as did almost all of Britain's national newspapers.

(Strangely, the US, which happily doles out stars for movies, is more reluctant with stage shows. *USA Today*, for example, gives movie reviews out of four stars, but no stars to stage plays.)

The backlash duly followed. In 2023, Time Out London announced that it was doing away with lowly one- and two-star reviews altogether, thus, you could say, making the whole system moot.

'PSA [Public Service Announcement] that's our last two-star review for a while,' Time Out announced. 'As of 1 May, Time Out will be operating a new review system (that's a bit like what we did in 2021 [i.e. during Covid]) wherein shows that would have previously received one-star and two-star will go unrated, with stars (three to five) – only given for "positive" reviews.'

This was ratings playing all nice, being used to prop up the industry being rated rather than to help out the reader (they'd done the same thing after Covid in a bid not to bash an ailing patient).

It is a feature of review systems that people just won't leave them alone, like jagging at an itchy mosquito bite even though you know it only makes it worse. Although they have conquered the world as a grade ladder, no one can agree how many rungs that ladder should have – nor, in fact, whether they like ladders at all. Remember Mariana Starke getting mocked in *John Bull* and then getting death threats? Star ratings have come under comparable scrutiny from their inception – so much so that you could say the ratings have been under continuous assessment themselves.

In the US, the *New York Times* is predictably haughty about theatre and TV reviews, eschewing stars all together. But restaurants are deemed fair game – except during Covid, when the *Times*' 'food sheriff' Pete Wells suspended the star system rating in order to not hurt any businesses that might struggle during the pandemic, before reinstating it again in 2022 (just like Time Out London did with theatre, before changing their minds *again*).

A cursory search of self-appointed literary critics, most (okay, all) of whom I've never heard of, reveals several articles attacking star ratings from both critics and consumers with titles like, 'Star Ratings Are Not a Valid Way to Assess a Book and They Need to Be Retired',[12] 'Why The Star Rating System Has Lost Its Meaning for Me'[13] and 'Why I Don't Like the Star Rating System'.[14] But then Goodreads and Amazon, the apex predators of the publishing ecosystem, have star ratings splattered over every page. The system hasn't lost all meaning for them.

Meanwhile, over on the internet, no one gives a flying flute what moustache-twiddling legacy media critics think. They've just forged on with an entire culture based around star ratings and consumer reviews. You don't need me to tell you where this has got us; just look in your inbox. If you've read, seen, bought, sampled, enquired or tested anything at all in the last five years, you will have been asked to rate it. The world is charged with pentadactyl hot takes.

★★★★

The final assault on the summit of commodified opinion was launched online. The invention of the internet – an instant information hoard, accessible by all – meant two things: whipcrack access to anything you could wish for and a similarly speedy means to comment on it.

When it came to product choice and remote shopping, if you couldn't see or try before you bought, you needed some kind of holding-hand before you clicked to purchase. On the supply side, exploding e-commerce meant that every Tom, Dick and Harry.com could set up a convincing webpage and, within minutes, become a retailer. Add in eBay (founded 1995), CraigsList (also 1995), Facebook Marketplace (2016), Gumtree (2000), Etsy (2005), Vinted (2008), Ali Express (2010) and Temu (2022), and everyone had become both a cottage industrialist and a bewildered punter at the same time.

How was the casual consumer to know who was reputable and who was a grifter? If it was cheaper at u-mug.com than at Selfridges or Macy's, did that mean the buyer should beware? Or was u-mug.com just a bona fide bargain (albeit with a duff name?) As the scams and the stories of scams became more numerous, star ratings and consumer reviews became a guard rail – a flawed one, but a necessary stabiliser nonetheless.

At the same time as allowing us to buy everything at home, the World Wide Web also meant we didn't have to *leave* the home. Thus, one primary forum for assessment, which we might call 'real life' – trying things on, meeting up with other people and telling them about what you'd tried and what you liked or disliked – was replaced with another: leaving a quick review.

The Covid-19 crisis added rocket fuel to the ratings game. When shops closed and government restrictions forced consumers to remain at home, e-commerce in the United States, which had already been expanding, experienced ten years' growth in the first *three months* of 2020.[15] In the UK there was a similar precipitous spike.[16] The volume of online ratings and reviews, the bird on the rhino, surged with it: between 40 and 80 per cent higher during the core pandemic months in 2020, compared with 2019. Covid sent the feedback loop into overdrive.[17]

In addition, this 'we' I'm talking about, those who don't have to leave home in order to peruse and buy whatever they want, has changed too. A younger generation of consumers, Gen Z (who come after Millennials and were born between 1997 and 2012) have more power than ever – and think differently. Look at the ad slogans that predominate at the moment. Patagonia, the outdoor gear brand, say, 'Buy Less. Demand More'. Oatly, the European non-dairy milk alternative, go as far as to claim 'We Exist To Make It Easier For People To Live Healthier Lives Without Recklessly Taxing The Planet's Resources In The Process'.

To a gnarled cynic like me, this is just marketing claptrap, but Ben Beck, assistant professor of marketing at Brigham Young University in Utah, says that 'Gen Z really loves this messaging'.

'Gen Z wants to be able to make decisions on their own,' he told me. 'They want to discover and, because of that, consumer behaviour is driving them to reviews even more. They want high-quality ratings and reviews, reviews that are thoughtful and review platforms that are doing thoughtful things to help them.'

Word of mouth, marketing's secret sauce, has thus gone digital and the five-star system has proved the perfect conduit for the internet age. It is quick, recognisable and transcends international barriers, just like the internet itself. It is clickable, like filling in a tick box or scraping off a scratchcard. You don't even have to type.

And when it comes to buying stuff on the internet, it is pretty much the only game in town: very nearly all customers look at ratings and star reviews when shopping online, with 96 per cent specifically seeking out the bad ones to better assess a business. That's up from 85 per cent in 2018.[18] This has all happened with freakish speed, driven by a whacking commercial imperative; four *trillion* dollars of global annual e-commerce is influenced in one way or another by online reviews.[19]

What it isn't is standardised. From the beginning, those doling out stars have struggled to achieve any sort of a level playing field. One site's three stars might be really quite good, but might be another's really quite poor; three Michelins is the tippity-top of all human existence while three stars in Google Reviews suggests something is mind-numbingly meh. This inconsistency, it should be said, has been a feature of star ratings from the beginning. It took the AA, mentioned briefly above, nearly a century to bulk all of their rosettes, laurels and tchotchkes for cafés, caravan parks and rickety B&Bs into a simple system.

It has also taken some time for five stars, as opposed to seven or three, to become the de facto standard. Even then there are still refuseniks. Metacritic and Rotten Tomatoes have settled on a percentage rating out of 100, presumably to reflect the fact that they're aggregators aiming to embrace nuance. Or just to look clever.

★★★★★

But why five? There are wormholes within wormholes you can go down looking at bimodal distributions, social choice theory,[20] behavioural science and detailed statistical analysis in order to find out why five is the best number of stars to choose from. Some people get very het up; others think it's because we have five digits on each hand.

At the very least, five is a number that runs deep in our psychology. Humans, studies have shown, can quickly and innately distinguish between one, two, three and four things, like stones. Show us more than four and we struggle.* The tally systems of the world all go up to five before repeating.[21]

My own take on why five stars has become the Goldilocks option (three too few; seven too many, five just right) is based on far too much time spent on online User Experience (UX)† forums (as well as on my

---

* You can try this with friends and some gravel, although be aware that passers-by will think you are mad.

† User Experience (UX) refers to the overall interaction, emotions and perceptions a person has when using a product, system or service. If it's making you really angry, it's probably UX's fault.

own user experience filling in far too many Before You Go feedback surveys). Broadly, it chimes with the thinking of a software architect called Danny Varod,[22] who I'll paraphrase here:

A scale with just *one choice* is stupid. Your only other 'choice' is not to engage at all. Television talent shows operate a single choice system and they are also stupid, as I'm sure we can all agree.

If you use *two choices* (like/don't like), then although you have at least created a scale you haven't left an option for 'fine, not that bothered'. Given that this is most people's response to most things, you're actually nudging them to not vote at all. Puff goes the system.

In a system of *three choices* (for instance, great, okay and pants), people might be unsure how to rank things they kinda liked but on reflection weren't that mad on. The same would apply for things they disliked just a smidgeon. This tool is too blunt.

A scale of *four choices* once again lands you in the even numbers snag: there's no middle option for the meh stuff (which we agree is most stuff).

But if you use a scale of *five choices*, you have all of the key categories (duffer*, a bit bad, middling, goodish and great) covered so you can differentiate between things you actually liked/disliked and things you only liked/disliked a bit. Five gives that sweet sense of real agency without the agony of undue complexity.

Why not seven or eleven or twenty-nine stars? Although these would offer up more information, once you start getting into the realm of what differentiates catastrophically bad from merely naff (and the same at the smiley end of the scale), you enter the realm of nuance and interpretation. It takes more thought and more time to alight on a judgement ('Is this one of the best – or worst – toothpastes/doctors/

---

* A word I love, originating in the mid-nineteenth century from Scots '*dowfart*' 'a dull or stupid person', from '*douf*' 'spiritless' (*OED*).

airport toilets I have ever seen or not?') and, as we have seen, the whole point of a star-rating system is brisk simplicity. While many sites, Amazon included, offer up a fudge in the form of half-stars or decimal ratings below the stars, it's the big one-to-five that predominates. UX experts seem to agree that, beyond five stars, you're better off not using stars at all, instead asking people to use actual words to express their views.[23]

In addition, there are commercial reasons as to why five stars has become the de facto standard: it looks good in imagery, it fits on a twelve-sheet poster just as well as on your smartwatch and you can see it from the other side of the road on the side of a bus. Too many stars requires too much time to take in and too much space to print. Too few doesn't say anything. Remember – the rating system exists as much to sell as it does to decide.

But really the proof, as so often, sits resplendent amidst the starry pudding; Google Reviews, Trustpilot, Feefo, Amazon and, yes, my newspaper TV reviews all use a quinary, one-to-five-star system. The free market, common sense and the simple human instinct to do what everyone else does has led us here, so that, in e-commerce at least, those five stars have become one of the catalysts for the global economy.

The upshot? In a society where everything – and, as we shall see, everyone – is rated and reviewed, the humble pointy glyph has become one of the dominant iconographies of the modern age, a global sign language that can encapsulate all kinds of meaning and sentiment in just five simple symbols. If the pen is mightier than the sword, then nowadays the star knocks pen, sword, reasoned debate and diverse opinions into a cocked hat. Behold the palm grease, the abacus, the bullwhip and the gilded lure of 21st-century social interaction:

★★★★★

Chapter Two

Why You Rate How You Rate

Today I bought a vacuum cleaner. It is a purple Shark IZ202UKT cordless stick vacuum cleaner with anti-hair wrap, up to 40 minutes runtime, pet tool, crevice tool and upholstery tool. It is the number-one bestseller at the time of writing on Amazon, with 2,209 ratings at an average of 4.5 stars.

I bought it while I was doing something else. I probably shouldn't have been flicking through product reviews at the same time as I was supposed to be concentrating on my work, but I suspect this kind of faulty-tasking (doing several things at once and none of them very well) is how many people spend much of their working life. The question – and I here offer up my own addled mind as a test case – is what is happening at a neuropsychological level when we engage with ratings and reviews in this way? How was I affected by what the world had to tell me about vacuum cleaners? Where do these behaviours come from?

Step one in any purchasing process is to identify a want or a need. Very often wanting and needing are the same thing. Very often they're inspired by clever marketing that's pressing subconscious buttons as much as anything else. I can't say for sure whether or not I was tacitly channelling some targeted vacuum advertising I'd seen last week when I decided to look for a new hoover, but it had been bothering me for some time that my garden office, where I am writing from, was a hole. And so to the frontline: which one to get?

I approached the task with relish – slightly more relish, truth be told, than actually getting on with the work I had to do. This is because choosing is fun[1] – even if you're not aware of it. The very act of choosing gives us a sense of agency and a feeling that we are in control of our own lives. It's well known in cognitive science that we like to impose narrative on our chaotic existence in order to provide an (illusory) sense of purpose and direction.* 'I am an intelligent being who looks at the research and then makes a measured call about which vacuum cleaner to buy' is a better story to tell yourself than 'I am a patsy for marketing departments and algorithms that separate me from my money like candy from a baby'.

Exercising control in this way, or feeling like we're exercising control, has been proven to be an innate need. In 1965, the psychologist Martin Seligman conducted a now-legendary series of experiments at Cornell University in which he gave dogs (harmless) electric shocks. From this he devised the theory of 'learned helpnessness' – in summary, how much control animals *think* they have is far more important than how much control they *actually* have.[2]

More than that, however, we also like the process of making choices, which is why having decided rather arbitrarily that I 'needed' a new vacuum cleaner, I set to the task of choosing one with great gusto. In one study, infants as young as four were shown to enjoy pulling a string to make music play; when the same music just played arbitrarily, they got angry.[3] Likewise, rats in a maze will choose to take

---

* As Joan Didion said in her book *The White Album*, 'we tell ourselves stories to live ... We interpret what we see, select the most workable of the multiple choices.' This is plainly a reference to star ratings 30 years before they became ubiquitous.

a meandering, convoluted path to food over a direct path.[4] Likewise, when I get a flyer for an exciting new vacuum cleaner, I put it straight in the bin. But when I get to spend hours picking my next vacuum cleaner, I'm on it like a vacuum cleaner on a pile of dust.

As the Columbia Business School psychologist Sheena Iyengar puts it, 'the power of choice is so great that it becomes not merely a means to an end but something intrinsically valuable and necessary.'[5]

These days, thanks to voluminous research and the widespread availability of fMRI (Functional Magnetic Resonance Imaging), we know where in the brain these choices take place. As I head to my laptop and start thinking about vacuum cleaners, that whirring sound is my prefrontal cortex clicking into action. This is like the bridge of the Starship Enterprise for the brain, the c-suite where your brain's executive board sits and takes decisions. Beneath the prefrontal cortex sits the striatum, which is in some ways the brain's first-line rating system. The striatum assesses the relationship between actions and rewards. When you scoff a pot of cookie dough ice cream, it is the striatum that tells you 'This is good,' (just as, when you stub your toe, it tells you the opposite). The striatum rates each experience and thus determines what you want.[6]

The thing is, what you want right now is not always the best choice in the long term, as any ice-cream fiend who now can't fit into their trousers will tell you. It's here that the prefrontal cortex goes to work. It is your checks-and-balances system, bringing common sense, rationality and judgement to bear on all the hot-headed, teenage-boy instincts of the striatum.*

---

* There have been vast amounts of published research on decision making and the neurobiology of behavioural science in the last few decades, with Daniel Kahneman and Amos Tversky's 2012 bestseller *Thinking, Fast and Slow* being the canonical layperson's summary.

Suffice to say that there is a lot of prefrontal activity going on when you access star ratings and reviews. You are imposing a level of control over impulse, evaluating all the available options and plumping for the best one. The prefrontal cortex is – among many other things – an inbuilt rating system.

Making good choices in this way is incredibly important, not only in vital fields like purchasing portable vacuum cleaners, but also in the trifling stuff like being a successful species. Choosing well is how we control our environment and, as human beings dominate planet Earth (for better or worse), you have to concede that we are quite good at it.

So no wonder I dropped the work I was supposed to be doing and flipped to some high-level hoover assessment with elation: it was beyond my control.

★

The first port of call was Google, which, until recently with the rise of ChatGPT, was the world's first port of call for almost everything. I typed 'Best vacuum cleaner for small room' and was served with a list of websites.

It's worth noting that, merely by searching in this way – the way that everyone searched until AI tried to change all that – we are already neck-deep in a world of ratings and rankings. The very act of googling, one that has earned a verb all to itself, is premised on rank. Google's PageRank system is what has elevated the company to global pre-eminence (and its founders, Sergey Brin and Larry Page, after whom PageRank is named, to stupendous wealth).

PageRank, in Google's words, 'capitalises on the uniquely democratic structure of the web by using its vast link structure as an organisational tool'.

What that means is that PageRank interprets a link from one

webpage to another as a vote, or a positive rating. The more links – or ratings – a page receives, the more prominence Google assigns to it. But Google is smarter than just that. It also rates, in a nanosecond, the 'importance' of the pages that made the link or, in our terms, gave the rating. Google is, therefore, rating the raters. And that's how it knows which pages to show me when I ask what's the best vacuum cleaner.[7]

Like every mug, I think of myself as definitely not a mug, so I skipped past the obviously cut-and-paste sites like ratemyportablevacuum.com because ... well, why? It was because I didn't trust them. And there, within seconds, we find ourselves at the ground zero of all ratings and reviews, which is trust.

Trust as a concept is silverfish slippery. You'll have been going through the process yourself these last few pages. Does this bloke actually know what he's talking about? Is that thing about the agency experiments on infants and rats true? Is Benji even a real name? Trust is the oil that lubricates almost every transaction, material or emotional, that we make. It is what star ratings and reviews are all about.

The word 'credit' comes from the Latin 'credo', meaning 'I believe'. British banknotes have 'I promise to pay the bearer on demand the sum of ...' printed on them because they are not actually currency in themselves but an undertaking to provide that amount of gold if called upon to do so. Hence 'credit' involves the belief that someone is good for the money and will pay you back – just as 'credibility' on a personal level, and almost all relationships thereafter, is founded on an understanding that a person can be trusted.

Trust is intrinsic to all human societies from the ground up. When caveman number one tells caveman number two there's a watering hole over there, does number two believe him or is caveman number one fobbing him off so that caveman two will get disembowelled by a sabre-tooth tiger, leaving caveman one to have more food for dinner?

If he is being fobbed off, he's dead and Darwin goes to work: having good instincts about who to trust is therefore passed down through successful genes. Social groupings – and thus survival – depends on knowing who's on your side.

Perhaps more surprisingly, trust is also intrinsic to successful capitalism. We think of *Wall Street*'s Gordon Gekko and robber barons and masters of the universe talking about zero-sum games and shafting suckers in cut-throat business deals for huge short-term profits. But while that's good for screenplays, for a long time before financialisation took over,* boring old honesty and probity was the path to profit.

Richard Tilly's essay 'Moral Standards and Business Behaviour in Nineteenth-Century Germany and Britain' shows how business transactions a century and a half ago were seen not as a one-off, fill-your-boots id-fest, but as building blocks for future transactions.[8] Trust was the mortar. It engendered word of mouth, repeat business and long-term prosperity. Even PT Barnum,† the greatest showman, the master Svengali and a multiple mugger of mugs, actually made his money by pioneering several ideas of what we would now call good customer service, including giving people their money's worth.[9] (Barnum's six-storey showhouse in New York contained more than half a million exhibits and yet a mere 25-cent ticket allowed people to stay there all day. This was in the mid-1850s, when weekly salaries were commonly between $4 and $5.) Trust, playing fair and being known as

---

* Meaning the increasing dominance of financial motives, markets and actors in the economy, leading to the financial sector growing larger and more complex than the 'real' sector of goods and services.

† Barnum (1810–1891) was the quintessential American showman who founded the Barnum & Bailey travelling circus and was renowned as a hoaxer. The well-worn adage 'There's a sucker born every minute' is often attributed to him – though, aptly for Barnum, there's no evidence that he ever actually said it.

the good guys has always been crucial to commerce. And so Google's now famous dictum 'Don't Be Evil' was already sound business advice in the 19th century.

In the modern world, trust has become even more complicated. The internet, with its dazzling speed and cloak of anonymity, has brought about more opportunities to be untrustworthy, and to profit from that duplicity, than ever before. We have all received an email from a Nigerian prince offering untold riches for a dubious click-through, and everyone has been warned of – and yet probably still fallen for – one of the many scams that occasionally bubble up on to the national news. They can be quite brazen and, at times, remarkably imaginative (as long as you're not the dupe). But in an era where you rarely see the 'item' – be it a holiday, vacuum cleaner or planned cosmetic surgery – before you hand over your credit, literally and metaphorically, how do you know what's the best choice, what's not so good and what's an outright lemon?

Well, you do your own research. In the last two decades, the so-called 'research phase' before purchasing has increased dramatically in duration and depth. It is no coincidence that the idea of 'doing your research' has come to prominence alongside the explosion in e-commerce. If you plot a chart of the use of the phrase 'Do your research' in published literature[10] against a chart of the growth of global e-commerce since 1995, they map one and the same. The internet has become the place where you buy and choose and do things at the same time as it has become where you look up what to buy, choose and do.

Before the internet, product research took the form of either trusting in your favourite bricks and mortars store to only stock the good stuff or asking someone you trusted to tell you what they'd bought and enjoyed (word of mouth). Department stores, which offered a wide selection of brands, relied on shoppers' faith in the products the store had selected. It was a faith not dissimilar to a religion; once shoppers had decided, for whatever reasons, on their chosen creed, they would accept the brands and products that store had bestowed upon them like a form of providence. If Sears or Debenhams or Fenwick or Harvey Nichols said this was the best item, the fashion du jour, then ours was what not to reason why. It was no coincidence that the architecture of so many of these stores resembled that of a cathedral, with Le Bon Marché in Paris, Selfridges in London and Macy's in New York City all monumental, temple-like spaces. These were places you went to be told what to buy – and you were proud to be seen there.

In the UK, shops like Marks & Spencer and John Lewis built up stalwart reputations among the middle-classes by offering rock-solid price and satisfaction guarantees. You could rely on them. From its beginnings in 1800, Marks & Spencer founded its business on trust, in both product quality and pricing. The very first M&S was a market stall in Leeds and even that had a slogan that fostered certitude: 'Don't Ask the Price, It's a Penny.' Brace buttons were sold with the guarantee 'once fixed, will never come off', while M&S fabric experts helped the government develop a minimum quality standard for rationed clothing during World War II.[11] The deal was that the customer should take M&S's word for it that their products were all five-star, but the underpinnings of that pledge were a series of frankly unbelievable returns guarantees.* The top-tier star rating in Marks & Spencer was therefore that the product was on Marks & Spencer's shelves at all.

Likewise John Lewis, a store that became known as a middle-class Mecca in the UK. The company's slogan for years, recently re-introduced, was 'Never knowingly undersold'. The unusual word

* In 2025, for example, M&S introduced a one-year quality guarantee on all children's wear: a full refund is offered within 365 days if the item does not meet quality standards.

'knowingly' asked customers to see the store as a sort of benevolent auntie who, though they might make a mistake and charge you more than another store for an item, would be mortified to find out that they had done so and quickly rectify the error. As with M&S, for years it was a running joke that you could take items that you hadn't even bought in John Lewis back to the store and they would do their best to replace them.

But the explosion in e-commerce has not been kind to either store, just as it has etiolated most so-called 'bricks and mortar' retailing. As you no longer needed to go into a shop at all, specialist expertise – Carphone Warehouse for mobile phones, PC World for computers and so on – and trust in a store brand were quickly superseded by price as the sole defining quality. You found out *what* to buy online by looking at ratings and consulting bulletin boards and social media. You then went to buy it at whichever online retailer sold it most cheaply. The explosion in price match guarantees came about because of online 'showrooming', where consumers would go into one of those retail cathedrals, look and touch and assess what they wanted to buy, and then go on to their phone while still wondering round the shop and order it on whichever e-retailer was selling it most cheaply.[12] Likewise, these days everyone has a copper-bottomed returns policy, because the internet is so hyper-competitive that if they don't, they're toast.

The coronavirus pandemic of 2020 compounded the ratings boom, confining us to our front rooms on lockdowns (or with illness) and thus even further removed from word of mouth and bricks and mortar stores – while even more reliant on the internet. As a result, e-commerce exploded yet again, and a furious rating and reviewing epidemic began. Everyone was now doing their own research. They had no need for decades-old department stores or trusted brands or even friends who knew about this stuff to do it for them.

But doing your own research is based on a fallacy. The ease of the internet or the magic of a quick ChatGPT prompt very often means you are actually using someone else's research to make you feel like you've done your research.

Whether we realise we're doing it or not, therefore, we tend to turn to trusted sources and assessors for information.[13] (And bear in mind that AI, in the form of Large Language Models at least, can only do the same; it scrapes millions of sources for its patina of expertise, but it is a scraper nonetheless.)

One source of information is advertising. You probably laughed at that last sentence because, after more than a century of what HG Wells called 'legalised lying', no one trusts advertising. Indeed, repeated Nielsen 'Trust in Advertising' surveys show that the title of their own survey is the best joke in it (it's also the only joke in it). When an influencer or an online ad tells us something is great, we cry bullshit. When I am repeatedly told that 'This pillowcase is quickly becoming the must-have holiday gift of the year', I know that it's not. Whereas, word of mouth – recommendations from someone you know – is trusted by nearly nine out of ten of us.[14]

Consumers are growing increasingly wary of the information they see online when it's explicitly linked to promotions and marketing. Of course they are. This is why genuine consumer ratings and reviews are so valuable (and why marketing departments work so hard to make promotions not look like promotions). Obviously, it depends on which reviews and ratings you go to, but having spent the time to look them up, you are far more likely to trust them than an outsized claim on an oversized banner ad.

A market is another source of aggregated information, which is why Amazon tells you what is its best seller. Price is another: the price of a service or product gives you some idea of what it's worth to buyers and sellers, following the laws of supply and demand. If it's cheap as chips, there's probably a reason for that. Stella Artois beer used to be marketed as 'Reassuringly expensive', which is precisely the sweet spot where fuzzy emotion and cold economics make whoopee.

Corporations themselves – brands, marques, even whole nations (think German cars) – used to inspire trust. There is less of this since the late '90s dotcom bubble and the 2007–8 global financial crisis proved that so many big, trusted institutions were built on sand and magical thinking. The institutions that were supposed to have curtailed the bankers' corruption turned out to have enabled it. We the people had to bail out the banks that had bet against we the people. And no one went to jail.

Aside from its many other consequences, this led to a general malaise, one of mistrust and, for consumers, a promise to ourselves not to get fooled again by people and businesses who sold themselves as copper-bottomed. It led to a search for certainty. After the great mask removal of big business in 2007–8, we trust only when we verify, to borrow the Cold War slogan. Happily, the growth of big tech, while supplying more vendors and chancers to mistrust, has also provided a way of facilitating this verification: star ratings and reviews.

★★

Let's go back to my vacuum cleaner. When I sized up *Good Housekeeping*, the *New York Times* and the despicably named RTINGS website, I found that several models seemed to keep being recommended. To get the bigger picture, I headed to Amazon, as most people do. (Amazon is the most common place to start the purchase process; 50 per cent of consumers search here first when shopping online.[15]) I typed 'vacuum cleaners' and was presented with 20 pages of what I was assured were the very best rug suckers. Before I looked for any details – price, size, type, colour – I went straight to their ratings, reviews and number of ratings.

This is what most people do. In the last few years, ratings and reviews have become the primary factor affecting purchase decisions, more important than price, free shipping, brand and recommendations from family and friends.[16] Ninety-five per cent of customers look at ratings and read reviews before buying something.[17] I hope the ones who don't are okay.

At this point, I had already spent several months researching this book, so I liked to think I was pretty savvy about the manifest defects in the ratings system. Firstly, perhaps conversely, the stars people give don't always represent what they actually feel. If that seems odd, think about someone rating a day out at the races – let's say, 'brilliant once you're in, huge queues beforehand'. Or a super-duper pair of long johns someone's just bought, which are the greatest long johns ever committed to legwear, but cost more than a small car. With just five stars to choose from (and even though, as we've discussed, this is the optimum amount of options in order to get the rater to say anything at all) there is no way that all sentiments can be covered with such a blunt tool.[18]

And then there's the fact that the system has been gamed. We'll come on to how in Chapter Eight, but suffice to say that plenty of reviews are fake. They're not hard to find. Here's one from my vacuum cleaner.

> ★★★★★ GOOD PRODUCT 'I am pleased with this vacuum cleaner. It is easy to use and works well on all surfaces. The charge lasts for an adequate time for my house. It lifts the dog hairs and doesn't get blocked like my old vacuum.'

Yeah, thanks AI. (Or apologies to 'Grace Anne Makin' if you do actually speak perfect robot.)

Amazon will tell you that they take great pains to blat the bots and take down fake reviews, but despite their best efforts, I know full well that a good portion of these reviews are bogus.

So why didn't I just discount the whole lot? It's because, at a certain level, it's impossible not to be blown along by sheer preponderance. You'll know the feeling: it's 'Yes, I know, yes, Chinese review bot farms, yes ... but *5,068* ratings?' There comes a point where the weight of numbers carries the day.

In a 1969 paper, the social psychologists Lawrence Berkowitz, Leonard Bickman and Stanley Milgram told of an experiment that will ring true to anyone who has indulged in what in the UK is known as rubbernecking.[19] They asked a man to stand on a street corner and stare at the sky. They then watched to see how many passing commuters would themselves stop to see what he was looking at.

When one man stared at the sky, hardly anyone stopped. However, when four of the psychologists' placemen stared at the sky, 20 people stopped. The more sky-starers Berkowitz, Bickman and Milgram put on the street corner, the more pedestrians stopped and gawped with them – all of them gawping at nothing.

Not realising that they had inadvertently also invented the flash mob, in their 1969 paper the conclusion was that this was an example of what they called 'social proof'. People were stopping and staring because they assumed that, with so many other people stopping and staring, they must have something worth staring at.[20] People are sheep not through simple conformity or peer pressure, but through a syllogistic form of reasoning, one which runs, 'Well, they wouldn't just be standing there staring at *nothing*, would they?'

When it comes to the number of ratings – now up to 5,070 in the case of my vacuum cleaner – it is this social proof that renders the numbers compelling. Even though we know full well that some of those ratings, possibly several thousands of them, will be fake or from marketing agencies or aggrieved competitors or just annoying berks, we are still affected by this virtual crowd staring mindlessly on a street corner. And so, assuming my Shark's 5,072 (up by two more now) ratings are significantly greater than other similar products, it will always carry some latent influence.

What about the ratings themselves? I only have to read down to the first one-star review to see that category errors are rife. Rater 'Stu47' is angry that his vacuum cleaner arrived late and so has, in his own words, 'knocked off a star'. Failures in delivery and packaging are obviously nothing to do with the product itself, yet this kind of error is commonplace. Some people rate everything badly because they think it's funny. (Sometimes it is quite funny.) Or they are just very angry today, perhaps because they have an irksome hangnail or their partner has left them for someone younger and more attractive. Amazon's algorithm doesn't remove the star rating Stu left from the aggregate just because it's irrelevant, so it's perfectly possible that the Shark's 4.1 could have been a 4.11 if people hadn't chosen to vent their spleen inappropriately.

Equally, many star ratings are meted out in haste and with little systematic thought. We'll come to how and why this is later in the chapter, but in the process of assessing these ratings, I know this implicitly.

'Consumers when they're ready to review, they go in expecting some of the negative reviews to just be cranky people,' says the wonderfully named Mike Lemon, a VP of legal affairs at the US's National Retail Federation who spoke to me for this book. 'It's not like you just read a review and you're like, "Oh that person just cussed out the owner ... but I trust them." You're applying the same sort of scepticism to reviewers as you are to the business.'

You know this too: star rating is one of those least worst systems. (Spoiler alert: so is democracy. Spoiler alert two: I think these things might be connected, but we'll come to that in Chapter Six.)

It follows that some of star rating's flaws are also its strengths. The internet is an environment not only suffused with information but dominated by *positive* information. Sites promote good reviews and good vibes about their product. But a decade of research into how reviews and ratings work has shown that people spend *more* time inspecting *less* favourable reviews. It actually makes sense: we quickly discount a deluge of five-star perfection because it doesn't tally with the world around us, where nothing is perfect (except power showers and Exotic Fruit Soleros) and anything that appears so is most likely a fraud.

As a result, many websites will show you both top reviews and several other less promising reviews for the same product. They aggregate it into a star rating where four, not five, tends to be most people's benchmark.[21,22]

Having a forum where anyone can leave a review is a good source of what economists like to call 'private information'. This means independent experience, analysis and thinking that, in the aggregate, leads to an informed bigger picture.

On the other hand, the very act of making all those independent experiences public decreases independence in the round. People see other people's rave reviews or one-star put-downs, and they lead to a collective bias. Yes, you want diverse perspectives, but you don't want sheep.

We also know, however, that large numbers of people – including idiots called Stu who get angry about the postal service and blame their vacuum cleaner – tend to get things right when it all comes out in the wash. How so?

In sociology and economics, it was long thought that large groups made large mistakes. But James Surowiecki's brilliant 2004 book *The Wisdom of Crowds* had a different idea, one that has turned his title into a popular phrase.

Surowiecki set out to prove one thing. 'If you put together a big enough and diverse enough group of people and ask them to make decisions affecting matters of general interest, that group's decisions will, over time, be intellectually superior to the isolated individual, no matter how smart or well informed he or she is.'[23]

In other words, one smarty-pants know-it-all rarely knows it all, but collectively it's different: a confederacy of dunces can come together like a giant, pulsating superbrain.

Surowiecki does concede what we all know: that sometimes collective intelligence leads only to collective idiocy, such as people rushing out to buy that repulsive Prime drink for their kids or hoarding toilet roll during the pandemic. But to him this is negative proof of what good crowd decision-making needs – decentralisation, independence and diversity, not bubble think. And that is precisely what ratings offer – or, at least, what they are supposed to.

You might have just bridled at the word 'independence'. Possibly also 'diversity'. Surely looking at ratings is acting dependently, not autonomously. And surely the people leaving ratings are about as diverse as 50 per cent of a chess board. Subscribing to the wisdom of crowds is what sheep do.

But being a sheep has also got a bad rap. I am here to defend sheep, or at least some of them. Social science defines two types of sheepishness, and one is better than the other. Blindly following the crowd leads down blind alleys. But if individuals copy the right people in the right way, what sociologists call 'intelligent imitation', the whole group can benefit.[24]

What is that right way? It involves trying something out for yourself and seeing the outcome. But it also involves seeing what happens to someone else who's already done the trying-out. For good decision-making on this basis, you need lots of people to have tried the thing out and several of them to put their head above the parapet and not follow the groupthink.[25] And if you read that back, you'll see we have just described the modern rating and review system.

★★★

This is all quite a lot. A point comes at which I begin to wonder if I can be bothered with any more vacuum cleaner selection at all. There is work to be done and this has already taken up far too much time, time during which I could have got down on my hands and knees with a dustpan and brush and a J-cloth. Amazon is offering me 923 options after a search for 'handheld vacuum cleaners' across more than 30 brands I have never heard of with names like 'Fockety' and 'Asperonic'. Is the effort worth the candle? Do I really need that vacuum cleaner? We are in deep.

The treacle round my figurative feet is what psychologists call cognitive load. You will know the feeling from the supermarket aisle where all you wanted was a bag of frozen peas and instead you yourself have frozen amid the superabundance of choice. It all gets too much. Your brain seizes up with decision paralysis (psychologists like coming up with new terminology). You take the best decision you can come up with at the time, rather than continuing the search, because you have things to do.

What happens next depends on whether you are a maximiser or a satisficer. I didn't invent the words and won't be held accountable for them, but they do seem to encompass how most of us approach a ratings and review trawl.

Andrew Howes is a professor of computer science at the University of Exeter, but he's an unusual computer scientist. He has spent most of his career studying quantum science, which you could define as computer models of how people think and make decisions. Howes is also the co-author of the paper 'Informing Decisions: How People Use Online Rating Information to Make Choices'.[26]

'Star ratings are partly about efficiency,' he says, speaking to me for this book. 'The social and summative nature of them means you can spend less time gathering information. I know that when I see 5,000 positive reviews, well, there's probably not really 5,000 real ones, but with all the available options in front of me, that's the best I've got to go on.'

Some of us wouldn't be happy with that. Perfectionists, nitpickers and people with too much time on their hands are what behavioural scientists call maximisers, incapable of settling for anything sub-optimal. (They're also the people quite likely to describe things as 'suboptimal' when they mean 'normal'.)

But according to the American cognitive psychologist Herbert A. Simon, the best type of decision maker is not a maximiser, but what he dubbed a satisficer. The classical economic models assumed that human beings are able to weigh up the pros and cons of every option rationally and with their own best interests at heart. It's a model that stands up to very little analysis in the real world – is that how you make choices? It's not how I do; I wish it was. But it has proved remarkably durable in the ivory towers of economics, miles from any retail park.

Simon, across several papers and publications, proposed that in fact that's not how people work.[27] He came up with a concept he called 'bounded rationality', which said that 1) we're not half as clever as we think we are, and 2) given that we're not computers and that weighing up multiple options involves a lot of effort, it might be better to be a 'satisficer' – someone who makes a choice the minute they discover an option that meets a certain threshold.

This tallies. In a society drowning in choice, we just want a rubber ring – that is what Barry Schwartz calls 'the paradox of choice' in his book of the same name. We want simplification. Indeed, it's been established that overthinking can lead to worse decision making (which, frankly, is just great to hear).[28]

What we're approaching via a brief stayover in academia is that sense of getting to your wits' end and wanting to do something else that I was feeling about my vacuum cleaner research. I can confirm, having pulled all of my hair out, that there comes a point when you realise you have spent far too much time learning about vacuum cleaners and you really need to get back to work/family/sleep. At this point, what the brain and the body are both looking for is clarity and ease. Take the cognitive load off.

Or to return to academia, here's Professor Andrew Howes again. 'A means to reduce the time effort of choice is to simplify the decision by increasing the discrimination between alternatives.'[29]

What you're looking for is something like an optician's red and green chart, so you can say once and for all that it's A or B and have done with the whole thing. The simplicity of the star-rating system increases 'discrimination between alternatives'. It gives you a solid reason to make that purchase and exit the 'yes, but which is better?' vortex that has, by now, taken up a good portion of your morning (and which, on reflection, probably won't matter in the end anyway).

Choosing a vacuum cleaner, or the best dog poop bags, or a book, or even the most comfortable underwear, is not a decision to be taken lightly. Sorry, strike that, it is. But we also rely increasingly on star ratings and reviews for what marketers call high-involvement decisions. A holiday, a knee surgeon, a construction company to smash up and rebuild your house or a life assurance plan are all complex, one-off and emotionally charged choices that demand reasoning, along with commitment of time and effort (although dog poop bags can certainly feel like a high-involvement decision when you're holding one that's suboptimal).[30]

High-involvement also equates to high price. People read more reviews the more expensive an item is, particularly when that item is new to them or is a brand they've never tried before.[31]

But, actually, in every case, what you're really trying to do is to dodge a duffer. More than find the best, you want to avoid the worst.

This is a central part of prospect theory, Daniel Kahneman and Amos Tversky's seminal contribution to behavioural psychology. In essence, prospect theory says that we assess gains and losses differently at a psychological level – and that we really, really, hate the losses.[32] Deliberation, therefore, is primarily a form of arse-covering. What you're actually doing when you scroll down through all those ratings and reviews is trying to make sure that you don't get sold a pup. In human psychology, that is more important than making sure that you choose the very best. Once again, it's the satisficers who are on the money, even if they have a stupid name.

We are forever on the look-out for material on bad experiences, more so than on good, because, says a *Review of General Psychology* paper, 'the self is more motivated to avoid bad self-definitions than to pursue good ones'.[33] Bad self-definitions mean things like being a patsy, a dupe or just an idiot. And avoiding being an idiot means, at the very least, checking a product, person or service's star ratings and reviews.

Cleverly, ratings sites have worked this out. Showing ratings at all – providing the forum there and then for people to whinge and gripe – is those ratings sites revealing an understanding of prospect theory (that people want and need to see bad reviews the most) as well as expressing trust. It shows that brands are on your side (they're not) and they want to help (if it helps them). It shows that they have thought about everything that has been written above and have attempted to engage with prospective consumers as more than just pliant economic units. As the advertising guru David Ogilvy said in the 1950s, 'the customer is not a moron. She's your wife.'[34]

Where, you're probably wondering, does me being married to Amazon leave the floor hygiene of my shed? Wither the Shark? Well, as you ask, it's here with me now. I got as far as Amazon's Best Seller and Best Choice (the latter a 2015 concoction that Amazon says is given to 'highly rated, well-priced products available to ship immediately'), was invited to Consider These Alternative Items, didn't, and then, in a bid to shed my hefty cognitive load and get back to writing this book, clicked Buy Now. The very next day, I was luxuriating in the benefits of the Shark's proprietary anti-hair wrap and up-to-40-minute run time.

Once I'd actually hoovered the floor, however, there was no need to hoover it again, at least for a while. This could mean only one thing: it was time to leave a rating of my own.

★★★★

When I say 'it was time to leave a rating', what I mean is I had no interest whatsoever in leaving a rating, but was worn down by more than seven entreaties in a single morning asking me to kindly rate my recent experience.

In this I am not normal. In 2016, although 82 per cent of Americans said they read online reviews at least some of the time, only about half of people said they left ratings. Even fewer wrote reviews. Only one in ten people reported 'nearly always' leaving a review.

But, come 2020, in the foothills of the pandemic, when online retail became the only retail, all that changed. A combination of spending more time at the rating portal – on screens – and virtually everyone being forced to take their shopping online meant that rating and reviewing went through the roof.

According to a survey that year of more than 10,000 respondents, 52 per cent submitted ratings and reviews multiple times per month, while 68 per cent of them only needed to be asked once to comply. Almost nine in ten said they rated something at least once every six months. The survey was taken from 'active shoppers across the United States who have opted into offers and discounts from retailers', suggesting that these might have been pro-sumers more engaged in the hamster wheel of modern retail than most. But, still, the numbers back up what you'd expect: lots and lots of us are rating and reviewing, and it seems to be as popular among Boomers as it is around the Millennials and across the various Gens.[35]

Star ratings go round and round on a feedback loop. (Yesterday, for example. I was asked to give a star rating ... for a feedback form. 'How would you rate your experience of this form? (Optional)'; the form had already asked me to rate a garden supplier. I liked that they were generous enough to make rating the rating form 'optional'. Otherwise, it's quite possible that my face would have imploded.)

On the Amazon site, you run your mouse over the empty stars and they fill like pints at the bar. Linger over five and the word 'excellent' appears, four is 'good', three is 'fair,' two 'poor' and one 'awful'. Given that Amazon proudly sells almost everything in existence, the whole world can be assigned to one of five simple slots, like sorting mail.

Why is everyone so keen for you to leave them a rating, given that it opens their product or brand up to public criticism when the game is already won – that is, they have already taken your money?

The first answer is the obvious one: because ratings and reviews have been proven to lead to even more sales. The body of consumer research proves that, as a Spiegel Research Center (SRC) report from 2021 summarises it: 'Online reviews have a significant and quantifiable impact on purchase decisions.'[36]

In addition, star ratings are by now so omnipresent that any brand or outlet not offering instantaneous consumer feedback looks like it has something to hide. More reviews signals quality. The sheer volume of responses also works as insulation, as ballast, against any bad reviews. And, finally, brands use the consumer data you generously provide for product development, customer research and hoarding metadata (who is buying, who is responding, who is engaging and who might be worth hassling with seven further emails in order to sow the seeds for a future purchase).

But the SRC report adds an important handful of caveats. 'The degree of that impact [of reviews on purchase decisions] depends on a number of factors, such as star ratings, the nature of the review content, the number of reviews, the price of the item, and the source of the review.'

This is fairly obvious. A few bad reviews are fine, as long as they're just some fizzling duds among many other confetti cannons. Thousands of detailed protestations from disgruntled, reliable reviewers, though, is not fine.

And so it comes down to who is leaving the ratings. Again, this tallies with what you'd expect. It takes a certain type of person to leave a rating. You know the type. As you're reading this book, you probably are the type. I know I am. 'Super-passionate' is the euphemism for we consumer crusaders. ('Time-rich WFH blowhard' would be another way of putting it.) Whichever way you define the type, though, the review sites would certainly like you to be more of the type, which is why they continually pester you to leave a review.

One way to stop all this is to give in, leave a review and, after that, hope that they'll leave you alone. And so, after a single floor-cleaning binge with my new Shark, I was back on Amazon firing off stars.

★★★★★

Compared to the triage that trawling through ratings requires, there is an entirely different dynamic at work when it comes to leaving your own assessment. Why, firstly, are you doing it at all? Several factors come into play.

Sometimes it's just because you've been bribed – the pill of leaving feedback is often sugared with the promise of a voucher or entry into a prize draw (that no one ever seems to win). In the case of Amazon's vast marketplace, it is against their terms of service to be compensated for a review, but it wasn't always that way. Amazon used to allow 'incentives' for product reviews 'in order to introduce new products to customers'.[37] That genie proved impossible to put back in its bottle. Now, sellers sneak between the semantic gap of what's 'compensation' and what's an 'incentive'. You certainly won't get paid for leaving reviews, but you still sometimes get chits with your products offering a gift certificate in return for a five-star rating. If you're lucky, you might get invited to the gravy train of leaving reviews in return for getting sent free stuff that is Amazon's elite reviewer reward programme Amazon Vine. (We'll look at how the system has been hacked, patched and generally dismembered in Chapter Eight and at Amazon Vine in Chapter Nine.)

Why else might you bother to rate a product or service? There is an element of civic duty involved, of doing the right thing, in as much as you benefitted (possibly) from everyone else's rating so now it's time to do your bit.

The American author Robert Wright, in his 1999 book *Nonzero: The Logic of Human Destiny*, suggests that this low-level piece of quid pro quo is part of our evolutionary development. We have learned over generations that we can all do better from trade and exchange (buying stuff) rather than it being the winner-takes-all, snoozers-are-losers game of standard capitalist theory.

The maths of a man called George Price also proved that the root of altruism is selfishness (although his discoveries sent him mad and he ended up giving all of his belongings away). More famously, Robert Trivers' concept of 'reciprocal altruism' explained why and when co-operating with others who do not share our genes might provide an evolutionary advantage.

The upshot is the same: whether or not it's to help yourself in the long run, it's still good to help others, on a rising-tide-floats-all-boats dictum.[38] Taken at face value, you might want to help the company who you deem to have done you a good service by making a good product. You might even want to help them with constructive criticism, pointing out faults in their products that they could improve.

In the 21st century, those companies listen. They can't afford not to.

'We spend lots of time directly interacting with our customers, helping them use our data,' says Tim Chilton, a managing consultant at the British mapping service Ordnance Survey. As Chilton describes it, ratings and reviews are a route into a more prolonged customer interaction.

'They tell us what they think, and they tell us a little and often, so we don't get a big shock once a year. We are working with these people on a weekly, day-to-day basis. That allows us to fine-tune our product roadmaps regularly. You talk to your customers and ask, "What do you want?" And it normally isn't what you thought they wanted.'[39]

This is all very Pollyanna-ish, but we don't all leave ratings to be nice. Some of us enjoy the petty tyrant's power trip of administering a vituperative sucker punch, an embittered one-star take-down born of disappointment, a feeling of being duped and the corresponding need to bite back. After a positive experience and having received free samples, a negative experience is the third most common reason for someone leaving a star rating and review.*[40] Sometimes, the second and third reasons are linked – people will leave a spiny stinker in order to elicit a response, and hopefully some kind of recompense, from the

* If this seems surprising, bear in mind that most of the available statistics are gathered by review sites like Yotpo, Trustpilot or PowerReviews. Although they all say negative factors are less common than positive ones, the numbers are always close. In addition, I might opt for the Mandy Rice-Davies defence: they would say that, wouldn't they?

brand. It's easier than getting through to customer service, if that even exists.

Once they've determined that they are going to respond at all, reviewers tend to fall into three categories. Some are looking for affirmation or a response, as above. Being asked for, and then leaving, a review (and then being told it has been read by x number of people or has been liked or got someone's back up or just noticed at all) provides validation. Amazon and Yelp both give stars or badges to those who post frequent reviews. Google will crown you a 'verified' reviewer if you send them a video laying out your credentials. And getting verified, acquiring an ersatz qualification and some letters after your name, feels good. It's the same ego trip that the internet has packaged as 'connection' (which actually means momentary access to the kind of community that the internet has all but obliterated).

Finally, rating something publicly and leaving a review is a form of influence tennis. You examined the rating and made the purchase. For that financial outlay, you have been granted a modicum of power, a chance to have the same effect on someone else that their rating had on you. People like to feel important. Go make 'em laugh, make 'em think, stick it to BigCorp.com with your withering take-down (of a vacuum cleaner).

A 2016 paper published in the *Journal of Service Research*[41] split consumers into 'powerful' and 'powerless', but if you've ever looked through ratings and reviews, you know the types already. Powerless consumers, the paper concluded based on three studies, 'are more likely to post positive reviews when the forum consensus is also positive'. In other words, they go with the crowd.

'Conversely, powerful consumers are more likely to post positive reviews when the overriding consensus is negative.' These are the mavericks who just use the goddam rulebook to smash down the door.

It won't surprise you to learn that there's little middle ground, but that's not necessarily a bad thing. In 1977, the American political scientist Charlan Nemeth set up a series of fake juries. Into each one she would plant someone with a dissenting opinion or divergent stance. What she found was that even when that stance was borderline nutso, and never going to convince any sane person, just having it there made the other jurors think harder about their own views. Dissent, as Nemeth's book, *In Defense of Troublemakers*, shows, is useful. Just one naysayer – or, in our terms, one oppositional star rating – makes other raters more thoughtful.[42]

Naturally, we'd all like to think that we plough our own furrow and are all powerful consumers. It sounds better, for a start. But a cursory look at any pool of ratings shows that in general we still can't avoid the effects of social proof we talked about earlier. It is well known that social media and the internet reward polarised opinions. Angry, disgruntled or completely delighted consumers offer better engagement and will post more than the blasé brigade, and that allows big tech to target them for advertising – which is still how big tech makes most of its money.

A lot of work has been done over the past few decades looking at how this polarisation starts. Yes, people tend to follow other people to one corner or another of the argumental boxing ring, but who defines the sides to be taken in the first place?

The hyper-prolific American law professor Cass Sunstein has spent much of his career studying polarisation and groupthink. In his book *Why Societies Need Dissent*, he begins with a simple question: 'Why and when, do people do what others do?'[43] His answer, confirmed through experimentation, is three-pronged:

- 'People who are confident and firm have particular influence, and they can lead otherwise identical groups in dramatically different directions.'

- 'People are extremely vulnerable to the unanimous views of others. A single dissenter or voice of sanity is likely to have a huge impact.'
- 'If people are, by our lights, from some kind of "out-group", they are far less likely to influence us, even on the simplest questions. And if people are part of a group to which we also belong, they are far more likely to influence us, on both easy and hard.'[44]

For our purposes, what that means is that the first ratings, and the first raters, set the tone. If you come to a pool of ratings to leave your own and the majority of ratings are five star, the chances are that you will be swayed in that direction. That's obvious. What kind of person starts the chain? The evidence suggests that it's the earliest, loudest people who are first to say their piece, which would be fine if they were the best informed. But guess what? They're not.[45]

It tends to be people who have ascribed themselves, or been ascribed, some kind of higher status who speak first and speak most. Often that means men, but online forums like Google Reviews and Amazon compound the problem by creating Verified Reviewers, giving them badges and rosettes, which fuels the fallacy that some people's opinions are worth more than others. In fact, merely being voluble, talkative, a chatty man has a major impact on how your opinions affect group dynamics, as Sunstein and the professor of behavioural science Reid Hastie have showed in an excellent (and excellently named) 2014 piece in the *Harvard Business Review*, 'Making Dumb Groups Smarter':

'Groups err for two main reasons. The first involves *informational signals*. Naturally enough, people learn from one another; the problem is that groups often go wrong when some members receive incorrect signals from other members. The second involves *reputational pressures*, which lead people to silence themselves or change their views in order to avoid some penalty – often, merely the disapproval of others.'[46]

Think about any forum of pooled opinion – say, for instance, the office party planning committee. Talkative people may not make themselves popular, but they do tend to be listened to.

This is self-perpetuating. People who post, review and respond to other's ratings tend to become more important in the discussion. Of course, there is no correlation either in the literature or in our personal experience between chatter and proficiency.[47] In fact, the gobshite who you wish would put a sock in it is generally protesting too much. Most reviews are posted within a week of receiving the product (76 per cent) or within the third usage (86 per cent) of the product. Few do so immediately, but a – surprisingly high – third of all consumers surveyed will provide a review after only using the product once.[48] In other words, they're rating something on very little evidence or experience. Like me with the vacuum cleaner.

On the other hand, in any group forum you need to take a position: strong advocates, either pro or anti, tend to acquire importance in group discussions. This is why leaving a rating makes you feel empowered.

Every time we take to the keyboard be it to leave a review or to post anything at all online, we're always projecting an extension of selfhood to the world, meaning a snapshot of how we'd like to be seen. Social psychology calls this self-enhancement and it's linked to self-esteem, which as we know is what allows most of us to get up in the morning (or, if we're lacking it, stay in bed).[49] For our purposes, it's enough to admit that you wouldn't post if you didn't want it to make some kind of mark somehow. This is why reviews from verified buyers are substantially more positive than reviews written by anonymous sources. (Conversely, anonymous buyers are significantly more likely to give one- and two-star reviews.)[50]

It's interesting – or, more accurately, dubious – that I feel qualified to opine on vacuum cleaners on the basis of having bought one of them. I am not, I am quite aware, an expert in the field. Before the internet, self-publishing, posting and blogging (and hence before rating culture took over the world), this might have stopped me from venturing an opinion. After all, what do I know? Why would anyone care what a TV reviewer thinks of a hoover?

But there has been a profound shift in the notion of what constitutes expertise and the extent to which we should bow to it. Firstly, this has come from an understanding drawn from politics and psephology – that experts only offer part of the picture.

As the American legal scholar Richard Posner says, 'experts constitute a distinct class in society, with values and perspectives that differ systematically from those of "ordinary" people. Without supposing that the man in the street has any penetrating insights denied the expert, or is immune from demagoguery, we may nevertheless think it reassuring that political power is shared between experts and no-experts rather than being a monopoly of the former.'[51]

The same goes for that one-to-five-star vote we're all casting all the time. If the point of a star rating is 'Tell me, what should I expect?', then we're into the realm of forecasting. No matter how clever we have become, no matter how many AI Cassandras there are out there, forecasters get it wrong all the time. Your weather apps prove as much most days of the week.

Fund managers, who are paid handsomely to tell people which companies are going to perform best, mostly get it wrong too. So do political commentators opining on the next election. Seers do not exist, though we continue to wish they did (leading to the American professor J. Scott Armstrong's famous 1980 'seer-sucker theory', which says, 'No matter how much evidence exists that seers do not exist, suckers will pay for the existence of seers.')[52]

Secondly, in the age of instantaneous, mass communication, hot takes, subscribing, likes and engagement, there is no such thing as a layperson. Your voice matters, whoever you are, however bigoted you may be or however little you've read. As much as it pains me to concede that there may be advantages to this age of stupid, the importance of everyone's opinion is not just feelgood marketing cobblers.

'Curation is valuable precisely because it avoids the algorithmic and the aggregative,' says the writer Michael Bhaskar in his 2016 book *Curation: The Power of Selection in a World of Excess*.[53] That is, my mistakes and lack of expertise, my quirks and Stu's stupidity are all valuable. The age of Big Data has yielded countless valuable insights, but it has also ironed out quirks and serried the world into statistical conformity. It is the humanity in the human doing the rating that matters. Cock-ups are contributions too.

Again, there's something 'wisdom of crowds-y' in this, going right back to our old friend Aristotle in the *Politics*: 'The many, though not individually wise, collectively may be wiser than any individual.'[54]

But it also feeds into consumer politics, where the customer is famously always right. It's axiomatic that a well-dealt-with complaint is the first step to a future purchase. The entirety of brand building is to do with trust, and trust comes as much from what a company does when a product fails as from how good that product was in the first place. The upshot of three decades of e-commerce, social media shaming and consumer pile-ons is that no customer can be ignored ever again. So rate your purchase, please do. We value your feedback (or, rather, we value the feeling that it gives you being heard, and thanks for the metadata).

Once again, this is probably all going a bit too deep for my vacuum cleaner. I have to say that the mighty Shark seems great,* revitalising my sorry rug in seconds, so I opt for four stars (because nobody believes

---

* I was not paid for this endorsement, although had I thought of it in advance I might have been.

five, see above), a few Shark gags in my review to make people like me ('the apex predator of the deep-pile seas') and note that it laughs at dog hair and dust with minimal impact on battery life. This is information that I hope might be of value to someone else, so I am doing my selfish altruism bit too.

At the same time, the Shark has received plenty of good reviews so I am now officially a powerless consumer, which was not what I wanted on my business card. Anyway, I post it. Within minutes, it emerges as a Verified Purchase and then, glory of glories, I am told that two people have found this helpful. I am vacuum man. I rate. I matter.

It reminds me just a little of the feeling you get when you mark your X on the ballot paper on election day. This is no coincidence, I think. There is a similar frisson of democracy, a scintilla of agency and of having an active influence on an indifferent outside world that comes with star ratings and reviews. It's a subject to which we will return in Chapter Five. At any rate, posting a quick-fire rating is one way of leaving the teensiest pockmark on the great, oblivious buttock of humanity. But when you leave a rating, you're also leaving a little bit of you out there. You are, in a sense, rating yourself. And corporations, retailers, marketeers, data miners and anyone with Wi-Fi now knows that bit more about you too. Happy vacuuming.

★★★★★

Chapter Three

Critical Mess – Star Ratings in the Arts

'I don't know where the stars come from,' the American film critic Roger Ebert once said, 'but they're absurd.'

Roger was right.* There is an absurdity to manhandling nuanced opinions and years of expertise until they fit into one of five star-shaped boxes. Part of that absurdity is that critics like Roger pretty much made their name using them; Ebert became one of the world's best-known film critics on the back of popularising and simplifying film analysis into his own 'two thumbs up' rating system. It is that paradox – coupled with just a soupçon of hypocrisy – that we shall explore in this chapter.

My day job – the one that you'll recall I should have been doing while I was choosing, using, rating and then gabbing on endlessly about a vacuum cleaner – is reviewing television programmes. Across several publications in the UK, I write about and then assign a star rating out of five to everything from big-ticket US dramas to one-off documentaries and homespun comedies. And I think I speak for many professional critics in saying that I am at best ambivalent about star rating in the arts.

---

* If he wanted to know where star ratings come from, I would happily send him a copy of this book – were he not dead.

Rating and reviewing art and culture, or in my case TV shows, comes with a different psychological tool kit to rating products, services or people. When judging what you like, an unbiased, impartial assessment is a contradiction in terms. As the late film theorist Victor Perkins (who hated star ratings) said, 'these judgements cannot be objective – but they must be rational'. I'm paid to be partial, but ideally I should at least make sense.

My own credo is 'firm but fair', which makes me sound like a Victorian nursemaid giving a child some vapours for the croup, but it has stood me in good stead. I know that making television is hard and that no one spends years trying to get a show on TV with the intention of serving up dross. I also review books for *Private Eye* magazine in the UK; now that here I am writing a book, I feel slightly different about dismissing out of hand the travails of first-time novelists and hacks like me.

But, for my own sanity, I have to play a longer game, too. If you dole out five stars like confetti at a wedding, then when a bona fide masterpiece comes along, there's nowhere left to go. On the other hand, trash-bashing for the lols cheapens the pen.

Carol Midgley is the TV critic for *The Times*. In March 2025, she was asked by a reader what her requirements are for a five-star rating.

'A lot of it is about gut instinct (and your own taste, obviously),' she said, 'but I suppose if there was a "test", it would be that when you see your friends in the pub, you say, "Oh my God, you have to watch this". It's about the sharpness of the writing and if the performances leave you thinking about it for days. Sometimes it's hard to decide between a four and a five (both are very good), but the difference is probably in originality and impact. So many shows are forgettable these days. The five-star ones never are.'[1]

Luckily, no one then asked her to remember all of the shows

to which she'd given five stars. I'm sure I have forgotten many unforgettables.

If Carol Midgley and my thoughts reveal anything, it's that there is no strict science to starring culture. Very few publications will say anything about what directives they work under, largely because they don't have any. The publications I write for have never given me a ratings bible. I searched far and wide for a published blueprint for rating popular culture and could only come up with this, from the US showbiz bible *Entertainment Weekly*.

'When we started *Entertainment Weekly*, we decided the magazine should be as useful, helpful, and informative as possible. Faced with a potentially confusing explosion of entertainment choices, we needed some way to measure them all and keep them straight. We didn't want the usual one-to-four (or five) stars; everybody uses them. Neither did we want to use thumbs, à la TV's Ebert and Siskel; thumbs give you no shades of gray – and it's not easy to typeset them. We settled on grades because we've all been to school and we all know what an "A+" or an "F" means. That's why we use grades in every section.'

Entertainment Weekly, which itself was spun out of a section in *People* magazine called 'Picks and Pans', allows its critics the luxury of a 13-point system, which they apply to TV shows, movies, books and music. They do not grade on a bell curve, they say ('An "A+" and an "F" will both be rare,' their guidelines add, 'but the grades don't necessarily average out to a "C"') and they judge *Citizen Kane* and *Rambo* according to their standing within their genre, not against one another. When pressed, they offered up a detailed list of what the grades all meant:

A+ A rare and wonderful treat. An ambitious effort that succeeds in practically every way. Almost sure to be on an annual 10 Best list.

- A Superb entertainment. An unequivocal recommendation.
- A– An 'A' with some noticeable flaw: one bad performance, a draggy beginning, a grating score.
- B+ Good, with something special that stands out – one performance, one song, one memorable scene. A solid recommendation.
- B Good, something you'll be glad you took in, even if it doesn't become a classic.
- B– A 'B' with some flaw, but still worth one's time.
- C+ Average, with a performance, a script, a song, or something that stands out.
- C Average. Not good, not bad. A critic's shrug.
- C– A 'C' with a noticeable flaw, something that sticks out. Not a recommendation, not a pan, just a bigger shrug.
- D+ Pretty bad, all around, with something that stands out for the good: a bad movie with one great scene, a bad album with one hummable song, a bad book with a neat beginning. This is not a recommendation, but that plus is a way to give credit when due.
- D A bad effort. Most really bad efforts are 'D's, not 'F's. A 'D' isn't ambitious and it isn't a success.
- D– A 'D' with something that is really awful about it. Sometimes, the minus is given to penalise entertainment for excessive violence or sexism.
- F An 'F' is given to something utterly without redemption. A 'D' is a warning. An 'F' is a condemnation. It is the critic's curse.

At the bottom of all this hand-wringing, they added, 'of course … the most important grade is yours'.

That was published in 1990, a pre-internet age since when that last line – that *the most important grade is yours* – has changed everything.

★

Before the digital revolution, your grade didn't matter. 'You' had no way of telling other people what your grade was. And even if you did, people deferred to critics and august publications when looking for guidance.

By 2026, however, *Entertainment Weekly*'s 'potentially confusing explosion of entertainment choices' had grown into a roiling firestorm of options. In 2009, 210 scripted TV shows aired on network and cable televisions in the US. By 2022, that was up to 600. On average, around 99,000 new tracks were uploaded to Spotify every day in 2024 – more than appeared in the entire calendar year of 1989. Film, supposedly a medium on the decline, still managed more than 5,000 releases in 2024, up from 1,525 in 1989.[2] The multitudinous, instantaneous means of watching films old and new has translated into more movie choices than ever. There's something almost gluttonous about the phrase 'on demand'.

In books, meanwhile, the huge rise in self-publishing has meant that alongside the 200,000 books that are published annually in the UK alone, there are now an additional 2.6 million offerings put out there by amateur authors for readers to try.[3] Self-evidently, no human being could ever get through a fraction of all this entertainment (and, if they tried, it would quickly cease to be entertaining).

Alongside all of that, the rise of fan culture that's fuelled and abetted by online forums and subreddits that are petri dishes for controversialists and hot takers means that there is very little line to be drawn between fan and critic. Fans like to rate. Fans are critics. Ergo, critics must rate.

Nonetheless, for the professional critic, the same problems remain that presented themselves to Mariana Starke in her travel guide (and led Oliver Herford to call rating things at all 'a sweatshop of twaddle').

Kristen Baldwin is *Entertainment Weekly*'s television critic. She has been working at the magazine since 1995 and is therefore, like me, a frontline worker in that star-rating sweatshop.

'The one thing I very rarely do is give a straight A,' she says, speaking to me for this book. 'I will give an A minus because I try to reserve A's for *The Wire* or *Pachinko* or things that are truly above and beyond excellent. If everyone's special, then no one's special, and then the grade is meaningless. So I don't go handing out A's willy-nilly.'

The problem I find when star rating is not so much at the top or bottom end of the scale (and let's be honest: lauding or lambasting are generally where the fun is to be had) but in the middle.

When I decide that *Surviving Black Hawk Down*, for example, a Netflix film I reviewed, is four rather than five stars, I'm consigning it to a bracket from which nuance instantly evaporates.* The temptation is always to regress towards the mean, which is three stars – neither a damnation nor a garland, just a large, padded fence to sit on.

The truth is that everything is not awesome, as *The Lego Movie* would have it, but then neither is everything awful. Therein lies the problem: both the laws of statistics and your general life experience will tell you that most television, most stuff, even most people are to be found somewhere in the region of 'okay'. What's that can of Coke

---

* The idea of 'bracketing' as a verb comes from the military in the mid-19th century, meaning, according to the *OED*, 'find the range of artillery fire by means of a bracket; fire or land short of the and over the target'. In other words, bracketing has always been about firing off potshots and seeing where they land.

like that I've just drunk? Fine. How do you feel about the performance of the bin men this morning? Eminently passable.

But that means that saying something is three stars is not really saying anything at all. It's like the response you get when you ask your teenager what school was like today. So you go up to four, or down to two, and then suddenly you're edging dangerously towards competence, even perfection, or slumping towards censure.

Here are some examples from my review library last year. Was the second series of *Vinnie Jones in the Country* really as good as the finale of the 17th series of *The Great British Bake Off*? Probably not. But it was better than *Stacey Solomon's Renovation Rescue*, that's for sure. Unfortunately for Vinnie, but fortunately for Stacey, they were all three starrers in my published reviews. Three stars contains multitudes. Too many multitudes. But what star ratings are supposed to provide is direction.

Certainly, half stars help, and many publications and websites have gone that way. The upscale music site Pitchfork rates albums out of ten, and then into tenths, which is both slightly pretentious and also cheating. If you look at their old rating key (long removed from their website, but telling), you'll see that a) the system was pretty arbitrary, and b) it actually boils down to – guess what? – five ratings, with little difference between 'Indispensable and Amazing', 'Exceptional and Very Good', 'Above Average to Mediocre', 'Below Average to Heard Worse' and 'Awful/Breaks New Ground for Terrible'.

- **10.0:** Indispensable, classic
- **9.5–9.9:** Spectacular
- **9.0–9.4:** Amazing
- **8.5–8.9:** Exceptional; will likely rank among writer's top ten albums of the year
- **8.0–8.4:** Very good

- **7.5–7.9:** Above average; enjoyable
- **7.0–7.4:** Not brilliant, but nice enough
- **6.0–6.9:** Has its moments, but isn't strong
- **5.0–5.9:** Mediocre; not good, but not awful
- **4.0–4.9:** Just below average; bad outweighs good by just a little bit
- **3.0–3.9:** Definitely below average, but a few redeeming qualities
- **2.0–2.9:** Heard worse, but still pretty bad
- **1.0–1.9:** Awful; not a single pleasant track
- **0.0–0.9:** Breaks new ground for terrible

Pitchfork's preposterous 14-point rating system

Hotels, which gave us the notion of five being the most stars visible in the firmament, now seem to add a star to the gilded imaginarium of the hyper-luxe every few years. (The Burj Al Arab in Dubai, the one on a man-made island and where Roger Federer and Andre Agassi once played tennis on the helipad, began life as a seven-star hotel, but is now listed on Tripadvisor as – hold on to your gold taps – an eight.)

Again, if the purpose of star reviews is easy clarification, then star-flation, adding new gradations and decimal points, only muddies the waters. Metacritic, another reviews aggregator for games, movies, TV shows and music, uses a number scale from 1–100. But when it ports my reviews from the *Telegraph*'s five-star system to its own percentile points, I seem to be regularly pronouncing television shows completely perfect – which obviously they aren't (except for *Hey Duggee*, but that's another book entirely).

And, anyway, half stars, numbers, fractions and exam grades just aren't as effective. Science says so. A 2025 study at the University of British Columbia conducted 12 experiments and concluded that we over-estimate fractional star ratings when the stars are pictograms (for example, in our minds we up ★★★★⯪ to ★★★★★) and yet we underestimate fractional Arabic numerals (that is, when we see a written number like a 3.5 we instinctively determine that it's more like a 3).[4]

The authors called the overestimation the 'visual-completion effect' and the underestimation of numerals the 'left-digit effect'. (This is the same psychological juju, the authors added, that leads retailers to price items at £3.99 rather than £4.00.)

'We evolved seeing three trees, three apples and three stones, so we can't immediately make intuitive sense of what 3.5 is,' Dr Deepak Sirwani, one of the co-authors of the study, told the UBC Sauder School of Business website. 'So the brain searches for a round number to base its estimation on, and for pictures the number that's available is rounded up. For numerals, the number that's available is rounded down.'[5]

Products whose reviews land on whole numbers, the study concluded, do not experience a bump up or down.*

★★

Perhaps because of all this confusion, there have been countless arguments at newspapers and websites over the years about whether they should give star ratings to reviews at all. In 2009, YouTube ditched a five-star system in favour of just likes and dislikes because everyone was voting one or five stars anyway.

'Seems like when it comes to ratings, it's pretty much all or nothing,' said YouTube's group product manager Shiva Rajaraman in

---

* Writing those last three paragraphs I have discovered that getting a half-star image when writing is extremely difficult. I am using Microsoft Word and I can't find a half-star symbol for love nor money. The *Telegraph* told me that their Content Management System can't cope with half stars either. This may be one more reason why people are reluctant to use them.

an official blog post at the time. 'Great videos prompt action; anything less prompts indifference. Thus, the ratings system is primarily being used as a seal of approval, not as an editorial indicator of what the community thinks about a video.'[6]

Culture has wrestled with similar misgivings ever since it started assigning stars. To take just one example, in 2014 the comedy website Chortle headlined a feature, 'Isn't it time the stars went out?'[7] It criticised not just star ratings, but the fact that they have acquired the verbal helpmeet of being 'awarded' (which editor Steve Bennett called 'such a terrible verb').

'Asking a reviewer "How many stars is it?",' said Bennett, 'is like asking most comics "Tell us a joke" – reducing all we strive to do to a simple format for those not really interested in how things work.'

That seems a bit harsh on people who rely on star ratings as a pointer, which is to say absolutely everyone. We can't all be uninterested in 'how things work'. And, anyway, what is a review 'striving to do' in the first place? Sometimes it's striving for the truth, or to capture the zeitgeist, or to help busy people decide what's worthy of their limited leisure time. At other times, it's striving to sell something or a narked reviewer is striving to get something off their over-strived chest.

Either way, Chortle's disdain for star ratings is not unusual in serious-minded journalism. The common critique of the commoner's critique is that star ratings are simplistic, philistine, mildly insulting and that everyone would much rather not use them at all. And they have a point: one major downside of the rated society is that with the whole world weighed and measured on the fingers of one hand, there's little room between the digits for ambivalence.

Michael Billington, the UK's longest-serving theatre critic until his retirement in 2019, described star ratings as 'abominations'. In an interview with the magazine the *Fence* in 2023, he said, 'I remember

[when] they started to come in. We had meetings as critics. We told our editors that we didn't like them. And they would say: "Oh! We have to because the public love them." Now they are part of the landscape.'[8]

There's a key point there: the public love them. Why might that be? In part it's because it gives us normies a heads-up and a reason to read on. This is particularly the case with one-star reviews.

'We've found traffic-wise that people really love bad reviews,' says Kristen Baldwin back at *Entertainment Weekly*. Often it's not just because readers are sadists; it's because a bad review yields better writing.

'It's because there's a passion behind it,' she adds. 'It's like, "I'm not just disappointed by this. I'm *mad* about how bad it was."'

Indeed, in late 2025, the *Guardian* in the UK ran a whole series pegged to its no-star review of the Kim Kardashian TV series *All's Fair*, a trashy legal drama. The paper's critic said that 'Kim Kardashian's world-historically horrendous TV show *All's Fair* has detonated a firestorm of critical horror which, as well as everything else, may have undermined the currency of the star-rating review economy.'[9]

That firestorm of critical horror made great copy, however. The *Guardian* managed to eke out four separate features about no-star reviews, including a list of all the no-stars they'd ever published (18)*

---

* For the masochists out there, they include the 2002 film *Boat Trip*; The Vines in concert in 2003; a 2004 musical about Oscar Wilde written by the former Radio 1 DJ Mike Read; *Menopause: The Musical* from 2007; the Valkee 2, a device that claimed to cure Seasonal Affective Disorder; the seventh season of *Love Island* in 2021; and *Buying London*, a 2024 UK-based spin-off of Netflix's *Selling Sunset* following a band of high-end British estate agents. Most recently, the *Guardian* published an apology for giving *Melania*, an Amazon documentary about the US First Lady, a one-star review. It should have been zero, they said.

and a chance for readers to 'share your zero-star disasters' as everyone piled into *Ishtar* and *Waterworld* for the umpteenth time.

Part of the problem with ratings for critics is that it stakes your colours to the mast. It becomes all too easy to then compare your response to that of other experts and conclude that you got it wrong, or are really stupid, or on the take.

'I will say that if I review something and I hate it,' says Kristen Baldwin, 'and then I go on Rotten Tomatoes and I'm like the only critic or person who hated it, I do feel a little like, "Was I wrong?" And I have to remind myself that there really is no right or wrong.'

Baldwin is both right and wrong about there being no right or wrong, because the entire purpose of ratings aggregators is, ultimately, to provide some certainty. We know, of course, that certainty is an illusion, just as we know that if we order something online that comes bestrewn with five-star ratings like rose petals, there remains the possibility that it will be rubbish. But it is the paradox of limitless choice – in commerce as much as in the arts – that the more we are offered, the more we need help in selection.

Some critics embrace this. Peter Bradshaw is the film critic for the *Guardian* and he, unusually, loves star ratings.

'I'm solely responsible for what I put above my reviews,' he told his fellow *Guardian* critic Marina Hyde, 'and they're based on my personal critical judgement as is expressed in the text beneath.'[10]

Bradshaw likes star ratings because he says they force him to take a clear position. Fancy-pants, self-certified highbrow critics with humanities degrees *love* ambiguity almost as much as they like quoting Walt Whitman about containing multitudes. They will 'on the one hand this, on the other hand that' (check my text – I've been doing it the whole book) until what they actually think about something is drowned in a sump of highfalutin' waffle and allusion.

Whacking a rating above the text and then making sure the text speaks to that rating at least forces you to make a call.

And, in the digital age, publishing is all about taking that position. In the last decade, star ratings have started appearing in headlines, not just nestled above or below the main copy. This means they are the first thing the viewer sees and a key factor in the all-important SEO rankings. Remember Google PageRank, the ratings system for the entire internet we discussed in Chapter Two? The star rating in a headline affects the page ranking. It is the very first thing both Google and human beings (less important) look for.

At the epoch-making Oasis comeback gigs in the UK in 2025, the star rating was listed above the masthead in *The Times*, the *Guardian* and the *Independent*. It was deemed more important than chancellor of the exchequer Rachel Reeves being unable to rule out autumn tax rises. This is why people love star ratings for television, art and culture: they have become our cultural route-finder, the first bearing we take when looking for orientation in a world of dizzying media superfluity. We are lost without them.

★★★

As we showed in Chapter One, this wasn't always the case. Star ratings are a relatively recent invention. Two things have happened to make them ubiquitous: one is that we have become used to products and services being rated with stars, and the other is that the arts have come to be seen as another product or service.

Naturally, critics don't like this; conceding that the Caravaggio exhibition at the National Gallery is a commodity like a bag of chips (or a vacuum cleaner) implies that their considered aperçus and devastating witticisms are just part of one big commercial game.

Well, it pains me to break it to fellow critics, but they *are* just part of one big commercial game. The rise and rise of e-commerce, where a review is never more than a few pixels away from a 'Buy Here' button, where newspapers provide links to bookshops on the same page as they review books, means that the whole world has become transactional.

In tandem with that, the role of the critic has changed, subtly but indelibly. If that role used to be 'to mediate intelligently and stylishly between a work and its audience, to educate and edify in an engaging and, preferably, entertaining way', as the *New Yorker*'s Daniel Mendelsohn describes it,[11] the critic invariably spoke down to the reader. He or she knew more, understood more and now, in these hallowed pages, would be good enough to sow the seeds of enlightenment on the dry ground of the masses.

Over the past 50 years, however, the most popular critics have found a way to speak on a level with their audience. The *New Yorker*'s own legendary film critic Pauline Kael pioneered an idiom that sounded like how people talked.* What's more, she spoke to them about the types of films they actually watched – blockbusters, popcorn movies, two-bit horror flicks. Criticism became a conversation between equals rather than a casting of pearls before swine.

Filtered down to the extreme, and supercharged by the internet, that conversation has become quinary. For the most part, the reason people read critics is to decide whether or not they should part with their money and see, say, *One Battle After Another* (yes) or pay £20 for the new James Patterson (maybe don't). Everything is a commodity and in the 21st century, when e-commerce has refined the sales-feedback superloop to a fine pitch, commodities are all rated one to five. That's what readers of reviews in newspapers and on websites expect.

---

* Like this on Cary Grant: 'Everyone thinks of him affectionately because he embodies what seems a happier time. We didn't want depth from him; we asked only that he be handsome and silky and make us laugh.'

But given that money is involved, there is more to this game than first meets the eye. It is open to abuse. In the case of television, critics receive what they call 'screeners' early so they can get first dibs. This access can be denied. Critics also receive 'care packages' and seasonal gifts, which are both euphemisms for bribes. There is, of course, no explicit command to leave a good review in return for your *White Lotus* Away hardcase carry-on or your Netflix *Squid Game* sofa throw. But studios aren't sending them out for the fun of it either. Motoring journalism, seasonal gift guides and pretty much anything that might lead to a purchase are almost all lubricated by this quiet greasing of wheels, courtesy of wealthy studios' PR agencies. It is, in this regard, no different to the Chinese tech firm you can't pronounce offering you vouchers in return for a five-star review on Amazon.

An insight into the grubby workings of star ratings came to light in 2013, with the release of the film *Despicable Me 2*. The *Telegraph* had agreed to a commercial package with Universal Studios in which they would run a series of tie-ins and advertorials written around the movie. Unfortunately, nobody mentioned this to the paper's longstanding (and excellent) critic Tim Robey, who promptly gave the film a two-star review. As reported in Alan Rusbridger's 2018 book *Breaking News: The Remaking of Journalism and Why It Matters Now*, this led to the commercial director storming into the conference meeting and the paper's then culture editor having a conversation with Robey in which it was gently suggested that perhaps the film was worthy of a three. Robey didn't change a word of his review, which on repeat reading does indeed read like a three. But the friction between the commercial and editorial arm of a supposedly impartial enterprise like a newspaper is telling.[12] And this was a decade ago. It is fair to assume that as long as film studios pay for advertising space both analogue and digital, there will be an element of self-censure in star ratings.

Happily, it works the other way with reviews submitted by the general public. A 2013 study by the economist Catherine Rampell in the *New York Times* found that 'critics regularly pan works that become immensely popular ... and praise some that roundly displease audiences'.[13]

Or, to put it another way, the public is much nicer than critics (except, weirdly, when it comes to black comedies and documentaries).* There's a fairly simple reason why: the public has actually paid for their ticket. They may have even wasted time queueing online to bag it, but one way or another, a selection bias comes into play, where you are predisposed to like the thing that you hoped to like. You go to a show wanting to have a good time. You handed over your money on the basis that you *thought* you would have a good time. To concede in public afterwards that it was not a good time but a waste of time makes you something of a muggins. 'It was everything I hoped it would be ... and more!!!' is a better line than 'Shouldn't have bothered with *Wicked* and now can't afford to buy food'.

★★★★

In books, the plot thickens. This is my first book and so I expect I am about to find out how ratings affect sales which then go on to affect ratings – I'll report back in Chapter Ten – but Richard Osman, the author whose Thursday Murder Club series has been one of the breakout publishing phenomena of the last decade, knows the ground well:

'Book reviews are not a huge deal,' he said on his *Rest is Entertainment* podcast in 2025, 'but book star ratings are *massive*.

* Rampell's Rotten Tomatoes study found that these were the only two genres where critics generally rate films more highly than the public. With romantic comedies, to take one example, it's the other way round. From this we can deduce that critics are glum and miserable.

The Amazon ratings, the Goodreads ratings and things like that are absolutely huge.'

It's not just the rating itself but, as we showed with our social proof study, the cumulative number of ratings too.

'People are slightly obsessive about them,' says Osman. 'And when I say people, I mean me. But it's fascinating on Amazon, just the way it works. Essentially, you think, *Gosh, if my book was four out of five, that would be pretty good*. Most books settle somewhere around 4.2, 4.3. And if you're above 4.5, that is like the greatest thing that's ever happened.'[14]

The strange thing, unique to books, is that they are widely pre-rated before your common-or-garden reader has a chance to read them. Go on to Goodreads, the world's largest site for readers and book recommendations, and you'll see that books that haven't even come out yet have somehow been mulled over and classified.

Right now, for example, I'm on Goodreads looking at Cal Flyn's *The Savage Landscape*, her follow-up to her excellent *Islands of Abandonment*. *The Savage Landscape* is, according to the blurb, 'a new book about our relationship to the natural world'.

'This book takes us into the wild – deep into dark forests, to the top of mountains and into the heart of deserts. It addresses our deep yearnings to be awed and inspired by landscapes that remain beyond our reach and examines what nature gets up to in the absence of humans.' As I feel that deep yearning to be awed quite deeply, particularly after a day in which I spent an hour and a half snarled up in a traffic jam on my way to pick up some water softener tablets, I go to buy *The Savage Landscape*.

The problem is that it is not yet published. And yet somehow it has already been rated on Goodreads, with its first review reading '1 three-star rating' (those were the words), which seems savage in its meaninglessness.

How could this be? Well, some of the reviews on Goodreads, it's apparent, aren't reviews at all. They're just saying stuff like how much they like the cover.

But, in the main, those who have pre-rated books like to boast that they have been able to do so because they have got a coveted ARC (advanced reading copy) from the publisher. Most of those ARCs, they then admit, came via something called NetGalley.

NetGalley is a members' website that shares reviews and, it says, 'champions literacy'. 'If you are a bookseller, book trade professional, educator, librarian, reviewer, journalist or in the media, NetGalley is for you!'*

The site exists to get the feedback ball rolling. If you are a committed reader, quite possibly one with a following on BookTok or Bookstagram, and you want to get books early, then NetGalley is a means to connect readers with publishers and let a thousand or so people have ARCs. The NetGalley-ian infantry then gets to fire off opinions, provide early feedback and, of course, give a star rating before the book comes out.

It's fairly obvious what's in the NetGalley transaction for those lucky enough to bag an ARC: they get the vicarious thrill of being a big player in the book world, and they can then show off about it to their followers.

The reason these early star ratings are vital to an author, however, is because the algorithmic recommendation system on which all e-commerce relies has to start somewhere. If you get some star ratings behind you, the Amazon algorithm will be kinder to you. So plenty of

---

* In publishing a galley means a book proof. The term comes from when the printing process was done by hand; a galley was a long tray that held a column of type. The typesetter would fill the galleys and then sent off a draft for approval.

publishers, particularly of fiction, will send them out to many people and will say to NetGalley reviewers, 'Please read it.' The only condition is that for being given the ARC, you must review the book and give it a star rating.*

Publishers are not stupid. They will never say that a condition of being sent an ARC is a four- or five-star review because that would quickly backfire. NetGalley, like any functional star rating system, relies on trust; if you read the submissions, early reviewers almost always mark that they were sent a book for free and that they intend to review it without fear or favour.

Indeed, having joined up myself a while ago for my book reviewing, it turns out that if publishers feel you're giving everything five out five, they might well stop sending you the ARCs.

Richard Osman again: 'It's a neat thing to get on to [NetGalley]. And the people who do it take it very seriously – you'll often see the earliest reviews are the longest reviews. Because they're "NetGalley people" who are used to doing this. You take that star rating very, very seriously.'

A lot of those 'NetGalley people' are influencers and literati, but many, I suspect, are already authors or wannabe authors themselves. Thus, NetGalley relies on a form of reciprocal altruism in which your own prospects as an individual (in Darwinian terms) are best served by everyone behaving honourably.

In this instance, it pays to be nice. The sociologist and polymath George Williams put it succinctly in his 1966 book *Adaptation and Natural Selection*. 'Simply stated, an individual who maximises his

---

* Many publishers, I should point out, use NetGalley in addition to the time-honoured route of sending out bound proofs to media contacts and review desks. It's also used predominantly for fiction.

friendships and minimises his antagonisms will have an evolutionary advantage, and selection should favour those characters that promote the optimisation of personal relationships.'

This is one reason why the internet as a whole tends to be suffused with positive ratings that, in the round, don't really reflect what many of us see as the overall mediocrity of the lived-in world. But if you love books, both reading them and writing them, then NetGalley is a salutary, generally supportive environment.

On Amazon and Goodreads, however, it's Sodom, Gomorrah and every stop in between ...

On Goodreads, 'the largest site for readers and book recommendations' that happens to be owned by Amazon, the largest site for buying books, your opinion contributes to a star rating too – but one honed down to hundredths of a point.

The competition is cut-throat. Small-timer William Shakespeare's *King Lear*, for example, clocks in with a 3.91 from 225,522 ratings. This drawn-out tale of a ranting geriatric with serious parenting issues has some catching-up to do on Liane Moriarty's enduring classic *Here One Moment*, which clocks in at a 4.10 (although with only 68,645 ratings). Admittedly, Moriarty's logline ('If you knew your future, would you try to fight fate?') is a much snappier sell than 'three hours of an old man shouting NO a lot'.

In recent years, Goodreads has taken a pummelling. The argument seems to be that it is not 'good'. A *New York Times* essay from 2023 – subtly entitled, 'Let's Rescue Book Lovers From This Online Hellscape'– summarised some of its problems. The issue, or issues, were that Goodreads – which, the author said, 'began in 2007 as a promising tool for readers, authors, booksellers and publishers' – had become something not dissimilar to the rest of the internet: 'an unmanageable, nearly unnavigable morass of unreliable data and unfettered ill will.'[15]

Several scandals – involving fake accounts, review-bombing and all the other suboptimal behaviours that dog internet review sites just as night follows day – had rendered Goodreads virtually unusable, the essay said. Nonetheless, looking at it now, nearly three years on, Goodreads is still very much used. For good or ill, any author not being chewed up and spat out on the site needs to ask themselves why.

On both Goodreads and Amazon, the star ratings always follow an interesting pattern. Early on, if you are J.K. Rowling or Bill Clinton or Jamie Oliver – someone famous who expects to sell a truckload of books – then your ratings are almost guaranteed to be woeful. Initial reviews flood in from people who know they don't like the author but who have almost definitely not read the book.

As the weeks go by, however, there comes a regression to the mean brought about by a larger sample size. If you're interested in buying the book, or just a sad-eyed stargazer like myself who's interested in how these things pan out, you should wait. Given time, you get to see what people actually think.

At the time of writing, for example, Tim Key, one of my favourite comedians and the star of the film *The Ballad of Wallis Island*, has just released a new book of poetry. On Amazon, it has just one customer review, from Hazel, and it's not a good one.

'The book is tiny and the font is so small that I am unable to read it. Having paid £14 for this item I am not happy. Why do these small authors feel it is acceptable to cut costs in this way. Disappointed with the author and the book.'

But a couple of weeks later and Key, not a huge name but a name, is clambering up the ladder. His average rating on Goodreads for his poetry is a 4.6. He'll get there.[16]

Goodreads was bought by Amazon in 2013. This completed the circle of reading, rating, recommending and selling, essentially making publishing a (drum roll, please) closed book. The Goodreads app and website are synced with Amazon.com for customer and reader reviews when customers use their Amazon account to log into Goodreads. Therefore, when a Goodreads member adds their review for a book on their Goodreads page, it will automatically be added to the reviews for the same book on Amazon.com. (However, the sync only works one way; if a customer leaves a book review on Amazon.com, the review will not show up on their Goodreads page.)

I think the take-out from all this is that Goodreads reviews are meant to be taken a little more seriously, and seen as slightly separate from the quagmire of Amazon reviews. There, all human commerce is rated (as we'll see in Chapter Four), whereas books are still granted something of a halo of superiority to, say, your preferred brand of baby wipe.

Is this kind of book reviewing criticism or commerce or something in between? It's a situation that would probably give legendary critics like F.R. Leavis or Susan Sontag palpitations, but it is where literary criticism is currently at. My personal contention is that, as so often with the rated society, we have found ourselves enjoying the least worst option. You could say that taking arts snobbery down a peg, with the proposition that everyone's opinion is valid as anyone else's, was long needed in arts writing. Professor John Carey, in his book *What Good Are the Arts?*, persuasively pricks the balloon of the sacred aura around the arts that has been built up over centuries 'to exclude certain types of people and certain types of activities'.

'How other people feel and think is,' Carey suggests, 'not precisely knowable. But most of us can describe our reactions to some degree, and an astonishing fact about the history of art criticism – including literary criticism – is that is has shown virtually no interest in this source of knowledge. It is standard practice for critics to assert how "we" feel in response to this or that artwork, when all they mean is how *they* feel.'[17]

Carey wrote this in 2005, a few years before consumer reviews and star ratings really took off, but he was correct: cultural elitism is rife in the notion of the professional critic whose views on the merits of a book or movie are somehow worth more than Johnny Punter's. Art and literary criticism have since been forced by the great democratising wrecking-ball of the internet to 'show interest' in 'how other people feel and think', as Carey put it. It's one reason why most reviews in newspapers now have star ratings in the headline, as we saw above; it's speaking to all the other lay reviewers, which is to say every reader, in a language we all understand.

But a by-product of everyone being a critic is that there is now a cacophony of cross-talk. Everyone has a voice to use if they want and, in star ratings, they all have a ready-made, globally recognisable insignia. It has both solved a problem and created another one: the sheer weight of published views about the sheer weight of newly released films and shows and books and TV shows has led to a renewed requirement for some kind of clarity. We are back to where we started.

★★★★★

This situation explains the rise of the aggregators. Review poolers like Rotten Tomatoes, Metacritic, IMDb and OpenCritic (in gaming) take both consumer and critic reviews as their raw data. They then feed the whole lot into their magical sausage machine and a number comes out. Metacritic, which ingests critics' reviews of TV shows, movies, games and music, spits out a figure between one and 100. Whether or not the critic gave a star rating, a figure, a grade or nothing at all, Metacritic assigns scores to their reviews and then the weighted average of those scores results in a single number: the Metascore.

Rotten Tomatoes, the pre-eminent rating site for film and TV shows, tries to have its tomatoes and eat them. Its most feted device is the Tomatometer, a yeah-or-nay icon that divides the screen world into 'fresh' tomatoes or a 'rotten' green splat emoticon.

It has been noted that this is quite a simplistic way to pack and stack the two pre-eminent mass modern art forms.

'What bothers me about Rotten Tomatoes,' says *Entertainment Weekly*'s Kristen Baldwin, 'is that it is so binary. Like, there's the tomato and there's the splat. I wish there was a third option – something like a slightly mushy tomato or a bruised tomato that you would use for sauce. Because there are so many shows that are B's or B-minuses that a splat feels mean ... but a full plump, ripe tomato feels inaccurate too.'

Critics would always like more rating options. Readers wouldn't. So Rotten Tomatoes tries to keep everyone happy: it also offers up a percentage rating, as well as a Popcornometer for an aggregate of its viewers ratings. But what about when it comes to how you leave your own rating? It's those five stars again. Rotten Tomatoes offers half-star gradations and, as you hover your mouse across the various options before posting your rating, it proffers helpful guidance (from half a star to one-and-a-half is 'Can't Recommend It'; two to three stars is 'Not bad but not my favourite'; three-and-a-half stars gets its own grading of 'Worth a Watch'; four and four-and-a-half are 'Definitely Recommend It'; and a five-star smash is an 'Absolute Must-See'). As always with star ratings, if you think hard about the difference between 'not bad' and 'worth a watch', you quickly realise there isn't one.

Why doesn't Rotten Tomatoes just let us pick our own splat or fresh ratings? Or give a number out of 100? It's because the five-star system is so deeply embedded in the collective consciousness, it's such a familiar, universal classification, that it's the one most people understand best. That's why newspapers and websites – who, if they are good at anything, it's knowing what their readers want – use star ratings.

Or they did. There has been something of a backlash, at least in the US where subscription models for newspapers have started to bear fruit. With another revenue stream rather than just clicks and page views, many of the country's more high-brow publications have freed themselves from what they see as the tourniquet of ratings culture. Their readers are soignée enough, they feel, to read reviews that contain nuance and no abrupt conclusions. (This doesn't seem to be the case in the UK, where star ratings still rule and we must therefore assume we are more stupid.)

The *New York Times* rightly points out that 'criticism doesn't stop with warnings and recommendations'. (The paper doesn't use stars, thumbs or letter grades, but especially worthy productions are designated Critic's Picks.) The *LA Times* and *Washington Post* don't use star ratings either (though the latter is owned by Amazon founder Jeff Bezos, so they've probably had it up to here with star ratings).

But whether a reader listens or cares what a critic has to say, it all comes down to the same quandary. In a world of limitless information and boundless commentary, who do you trust?

The *New York Times*'s co-chief film critic is A.O. Scott, also the author of *Better Living Through Criticism*. 'No reader will agree with a critic all the time,' he says, 'and no critic requires obedience or assent from readers. What we do hope for is trust. We try to earn it through the quality of our writing and the clarity of our thought, and by telling the truth. The truth, in this case, about what we thought about what we saw, read or heard.'[18]

Honesty remains the best policy. Unfortunately, the internet is a largely anonymous, amorphous, faceless beast where nuance goes to die. It's also the forum for almost all human (and, with AI, beyond human) interaction in the 21st century. So how do you trust someone in the digital age? And, in turn, how do you let it be known that your voice can be trusted?

★★★★★

Chapter Four

But Who to Trust?

Amazon is the everything store. I've just been on the site and today, as every day – particularly when it's Prime Day, or in the lead-up to Prime Day – its home page was a cornucopian, bewildering collage. In the shop window, there was indeed 'everything', from small dog daily dental sticks (4.6 stars, 346 reviews) to around a million different must-buy vacuum cleaners to go with the one I bought in Chapter Two (still going well, ★★★★). It's easier to list the things that Amazon doesn't sell than the things that it does. You can, and quite possibly do, get anything you could ever need to subsist from this one e-commerce leviathan.

The broad story of Jeff Bezos's brainchild is well-known and has been well told elsewhere, but star ratings and consumer reviews are a pillar of the company's history. The site launched as a bookstore and took its first order in April 1995. Two months later, a programmer named Shel Kaphan, one of the company's first employees, added a user-generated reviews feature, with star rating summaries following in 1997.

As we've seen in previous chapters, using star ratings wasn't Amazon's idea, not by at least a century. But having a shopfront where you concede that some of the things you're selling aren't that good was a radical step. Bezos saw that this was a numbers game. If Amazon.com could amass more user-generated reviews than anyone else, they would outflank other online bookshops and become the go-to destination for booklovers. Star ratings offered a quick way for readers to scan reviews and gauge a product's quality without having to read all the user-generated waffle.

But if you offer people the ease of a one-to-five-star scale, you're going to get some grouching. Some people *really* hate the books they hate.

According to his biographer Brad Stone, Bezos often said that publishers, who he needed to keep onside when Amazon was a bookstore, originally disliked the notion of customer reviews. Their fear, understandably, was that 'harsh and often anonymous online critiques could hurt sales'.[1] Bezos later recalled getting an angry letter from an executive at a book publisher implying that the Amazon man didn't understand that his business was to sell books, not trash them.

'We saw it very differently,' Bezos said later. 'When I read that letter, I thought, *We don't make money when we sell things. We make money when we help customers make purchase decisions.*'[2]

The idea that retailers were as much about offering direction as shifting units is something of a eureka moment for the rated society in which we now live. Where companies once existed to sell you stuff, using advertising and marketing to nudge or strong-arm you into that magic purchase, Amazon set itself up as both a shop window and a conduit for word of mouth. It is happy to tell you *not* to buy something if that means you'll buy something else on its site that's better. For Amazon – and thus all of the e-commerce tsunami that has followed – a one-star is as valuable as a fiver.

The other thing that Amazon realised was that it wasn't that important where the reviews came from – early employees and their friends wrote many of the initial ones themselves – but you did need a kernel of ratings for the system to function. To get that ball rolling, Shel Kaphan himself took a book off the virtual shelf that was about to go out to a customer, a riveting-sounding Chinese memoir called

Bitter Winds: A Memoir of My Years in China's Gulag. He read it at pace and wrote one of the first reviews.*

Allowing negative reviews garners trust. It is using other's bile to fuel your own likeability, in the same way that a good friend is the one who will tell you that your new boyfriend or girlfriend is a total loser. What you sacrifice in immediate sales ('Don't buy this book'), you gain in later sales ('but maybe do buy this one, and here's a one-click link to do so'). You are working with a global currency that is not dollars or pounds: it's trust.

What trust means in retail terms was best described by the psychologist Robert Cialdini in his classic book *Influence*. In the chapter entitled 'Authority', Cialdini tells the story of a waiter named Vincent. Over several meals, Cialdini noticed Vincent 'arguing against his own interests' when it came to recommending menu options or wines.

'When it was time for the first person, normally a lady, to order,' writes Cialdini, '[Vincent] went into his act. No matter what she selected, Vincent reacted identically. He leaned conspiratorially towards the table to report for all to hear "I'm afraid that is not as good tonight as it normally is. Might I recommend instead the–", where Vincent would recommend a pair of menu items that were less expensive than the dish the patron had selected initially.'

Cialdini's waiter 'seemed to have the customers' best interests at heart. To all appearances he was at once knowledgeable and honest, a combination that gave him great credibility. Who, after all, is more believable than a demonstrated expert of proven sincerity?'[3]

Becoming a demonstrated expert of proven sincerity would have

---

* Kaphan's review was published anonymously under the name 'A Customer', according to a *Business Insider* article from 2018. It was three words long and read, 'I love it'. It's no longer on Amazon, although *Bitter Winds* is, rated 4.2 stars and with 17 reviews.

got Vincent more tips in the long run, as well as making customers feel more warmly towards the restaurant he worked at and hence likely to return. Vincent was playing the long game: becoming a 'demonstrated expert of proven sincerity' is also the entire goal of the rating and review system, as well as the backbone of Amazon's business model.

'If vendors or brands leave Amazon, they will eventually come back,' said Jeff Wilke, Amazon's original head of operations, in 1999, because 'customers trust Amazon to be great providers of information and customer reviews about a vast selection of products. If you have customers ready to buy, and if you have a chance to tell them about your product, what brand ultimately doesn't want that?'[4]

Reviews and ratings are so important to Amazon that it is one of the few places where anyone can review anything that's on the site, even if they didn't buy it through Amazon. You shouldn't – but you can – review things you haven't bought at all. It is the lofty pinnacle of rating and review culture, where reviews of any hue are more important than anything else.

Amazon's definition for how it compiles its overall ratings is a classic of Silicon Valley obfuscation disguised as clarity. 'Amazon uses advanced models to calculate star ratings, not just a simple average. The system considers important factors like how recent the review is and whether it came from someone who actually bought the product. We also have safeguards to ensure reviews are authentic. The goal is simple: to show ratings that honestly reflect what customers think about a product. These ratings aren't fixed – they update as new reviews come in and older ones become less important.'[5]

It's interesting that a recent rating carries more weight than an older one. There is no definition of 'recent' given. You can see why older opinions might be less important than newer ones for outdated products, but surely not for books or films? Perhaps Amazon's model is so advanced that we mere mortals could never understand.

Or perhaps they're just reserving the right to diddle the ratings as they please, even as they insist that they are 'authentic'.

The simple goal behind all of this, according to Amazon, is so that customers can make 'informed, confident purchasing decisions'. The real goal, though, is more complex. When you click on that star rating or file that review, you're giving Amazon not just fuel for their own brand trustworthiness, but also entrusting them with details about yourself and your tastes. Among many other things, Amazon happens also to own the world's largest ad platform. Advertisers can use the Amazon Ads service to place ads on Netflix, Warner Bros. Discovery (HBO Max), Disney (Disney+) and NBCUniversal (Peacock).[6]

By filing a star rating, therefore, you are adding to Amazon's vast sump of commerce data – what we've all bought from them – to make sure we're served targeted ads that will appeal to us on TV. This means that Amazon is now the gatekeeper not only for the products you buy, but the products you're offered elsewhere.

The other goal for Amazon's ratings is to sell more stuff on Amazon. Across all businesses, there has always been a direct correlation between number of online ratings, reviews and volume of purchases, as we saw in Chapter Two.

Amazon has never really been a bricks-and-mortar retailer,* but its approach online has set the template for others to follow. It took Walmart, the US retail giant, ten years to follow Amazon's lead and add customer star ratings and reviews for its many products, which it started to do in 2007.[7]

'This is kind of a no-brainer for Walmart,' said Vikram Sehgal, research director at Jupiter Research, at the time. 'We've surveyed

* Amazon has dabbled with Amazon Go and Amazon Fresh convenience and food stores but closed them all in January 2026. It does now own Whole Foods, however.

consumers, and overall, they do find reviews useful.' In 2006, 48 per cent of those researching or buying products online described reviews as either 'very useful or useful', he said, 'and they often read through reviews to confirm that they are making the right decision. The whole idea is that there is wisdom of the masses.'

But, tellingly, Walmart was also looking for a dose of that precious trust capital from its customers. The year before it started posting consumer ratings, the company had been hit by a faux-blogging scandal that had damaged customer perceptions.* In the same month as star ratings first appeared on the Walmart website, it so happened that *Harry Potter and the Deathly Hallows*, the seventh and final novel in the mega-selling kid wizard franchise, was being released. Walmart was on a charm offensive. It asked its employees and customers to sign a pledge not to reveal the ending of the book, as well as handing out 50,000 sets of earplugs so people wouldn't hear who died (or didn't; I'm not saying) by accident. It was all part of a PR drive to assure customers that this faceless Big Corp was their pal. Walmart, like Amazon, could be trusted.

In 2024, Walmart conducted a survey of its customers to find out how its adoption of ratings over the previous decade and a half had affected consumer attitudes. The results showed that what Amazon had begun – and what the whole e-commerce world, right up to and including the world's biggest retailer Walmart itself, had continued – was now as much a cornerstone of retail as piling it high and selling it cheap.

---

* The faux-blogging scandal was an early example of what we now call 'astroturfing'. A blog titled 'Wal-Marting Across America' appeared online, documenting a journey by a nice-sounding couple named Jim and Laura, who travelled in an RV, stayed in Walmart parking lots and met with employees who really loved their jobs. The blog turned out to have been created by Walmart's PR firm.

'When choosing a product amongst multiple options, Walmart customers consider high rating or a high number of positive reviews a strong indicator of customer satisfaction with a product – product reviews are the #1 source of information that inspires their belief in a product's quality. They can also help build customer trust – allowing your brand to stand out, connect authentically, and build credibility.'[8]

Walmart was so struck by what they found from their survey that they published a blogpost for their Walmart Connect customers* entitled 'The hype is real: Building brand trust through positive product ratings.'

Highlights included stressing both quality ('Of Walmart customers surveyed, 51% say an ad with a star rating would inspire them to purchase the displayed product. High ratings help validate the quality of your product and, most importantly, show happy customers') and quantity ('It's not just quality – quantity is important too. 62% of customers say an ad with a star rating and high volume of reviews would inspire them to purchase. Showcasing a high volume of ratings in your ad creative can help move customers from considering your product to making the purchase. This creates a sense of credibility and showcases a product's popularity, building trust in your brand and what your product has to offer.')

Put them both together and you have the now-classic formula of an overall star rating plus a whacking great bulk bag of reviews. 'The last approach is combining both the rating and high volume of positive reviews. 77% of customers say that reviews inspire their belief in

---

* Walmart Connect is Walmart's own advertising platform. It allows advertisers to tap in to Walmart's 'vast omnichannel presence, the power of its closed-loop system and knowledge of its customer base to bring brands measurably closer to their intended audience no matter how they shop'.

a product's quality, while 70% of customers say that product ratings inspire their belief in a product's quality.'[9]

Parse those last three paragraphs and the words that stand out are 'validate', 'credibility', 'building trust' and 'belief'. For the world's largest retailer, ratings are a badge of faith in an increasingly fickle, bamboozling, multifarious consumer world.

It all points back to a truth that one Jeff Bezos noted in 2010.

'When you are 80 years old, and in a quiet moment of reflection narrating for only yourself the most personal version of your life story, the telling that will be most compact and meaningful will be the series of choices you have made. In the end, we are our choices.'[10]

This was a noble attempt at lyricism from a business titan, as well as an unacknowledged rehash of thoughts better expressed by Jean-Paul Sartre ('Nous sommes nos choix'), William James ('Selection is the very keel on which our mental ship is built') and even Marcus Aurelius ('You have power over your own mind – not outside events'). But it spoke to a simple verity. In our every decision, we want to make the best choice, the right choice, the choice that proves we are persons of good judgement and honed wisdom. We also don't want to be mugged off or played.

Star ratings are a public reflection on those choices, a rating of our own choices if you like. It's one reason why the internet is suffused primarily with four- and five-star ratings. A one- or no-star rating is an admission that you made a mistake, didn't do your research and ignored the old proverb that 'the obvious choice is usually a quick regret' – whereas the internet is a world of perfection and infallibility where everyone is living their best lives and choosing wisely. Your life should at least appear to be festooned with five-star choices, as befits your own five-star existence.

★

But the internet is also a brutal democracy and, as much as your Amazons and Walmarts might try, no one person gets to choose who or what gets rated. And so, in turn, retailers like Amazon have come to be rated themselves. In the past twenty years, the Danish site Trustpilot has published more than 300 million reviews of more than 546,000 brands.

What Amazon is for products, Trustpilot is for businesses and services. The company actually formed as the result of dodgy plumbing. It was 2006 and a nerdy 22-year-old Dane called Peter Holten Mühlmann was running his own e-commerce store selling electronics from his parents' garage.[11]

'A number of things were evident to me,' he says. 'A lot of people had very bad shopping experiences on the internet at large. On the seller side, there were huge trust issues. So often sellers would get a lot of visitors but not a lot of customers.'

What tipped Peter over the edge was a story he read about a travel agent who had sent a group of guests to a hotel where it turned out that the toilets weren't working. They were not offered another hotel. One can only imagine the scenes.

This incensed young Peter.

'I thought, you know I would like to have known that before I had booked!'

It came at a time when e-commerce and e-travel were well-established but not yet universal, particularly for your mum and dad: boomers might well still look for hotel recommendations in the newspapers or by asking a friend. Peter still lived with his mum and dad. He did not want to see them get caught up in some foreign blocked toilet fandango.

'So initially it was like ... hey ... there has to be a way where I can make it easier for Mom to get a good experience'.[12]

And so Peter decided that it was time 'to give people a way to express their honest, uncensored thoughts about a product'.[13]

He started a company that he called Trustpilot. His timing was shrewd and his acumen sharp: fast-forward 20 years and Trustpilot is the level boss of the ratings aggregator website world, rating every kind of business from hotels to ad firms to plumbers to plasterers from one to five stars. Trustpilot claims 64 million monthly active users across the globe, 140 billion annual 'Trustbox impressions' (I like to think that the word Trustbox stems back to Mühlmann's initial toilet trauma, although there is no evidence of this) and has offices in Amsterdam, Denver, Edinburgh, Hamburg, London, Melbourne, Milan and New York. Since you started reading that last sentence, 44 more reviews have been posted, consisting of a total of just over 150 stars, spread liberally like discarded glitter on a toddler's craft table.

They even have a star as their defining symbol, nestled in front of the capital T of their logo, and stars twinkle all around the Trustpilot homepage like lights on a Christmas tree. They revolve in animations and are subtly embossed with a small shading beneath the right-hand prongs that makes them stand out from the page. (The financial ratings aggregator Defaqto has the same logo in reverse.) Trust and stars, in this universal repository of belief, are complementary and interchangeable.

A little lower down the Trustpilot homepage, we get to see how the starred society is working out. If you want a snapshot of how civilisation is sallying forth in the 21st century, you could do worse than look at Trustpilot's 'recent reviews' shop window.

It tells us that at the time of writing a user named Asif is very unhappy with Booking.com. That Adam had a bad experience with The Unique Property Investment Group. Julian had an 'indescribable' experience using songer.co for his salsa project (one that he nonetheless feels able to describe at some length). Salman is annoyed that AliExpress doesn't ship to Pakistan. Shalini says that the headboard on a Koala bed wobbles. And Unified Mentor are great employers, which Harsh feels happy to declare despite his internship having been remote.

And so on.

Notably, all of the foregrounded reviews are either one- or five-star. Cherry-picked using some opaque algorithm, they tell you both quite a lot and also nothing (other than that life is like a box of chocolates and people are funny).

According to Peter Holten Mühlmann, Trustpilot was founded on principles of optimism and human possibility. Rousseau would have given him five stars all the way.

'When I started Trustpilot, a lot of people would say "Isn't it all going to be negative reviews?"' Holten Mühlmann recalls. '"Why would you bother to write a positive review?" Everybody suspects that everyone else is a total a***hole.

'It turns out the average star rating is 4.3 or 4.4 – people are nicer than you suspect.'[14]

It's a cheery idea in a mixed-up world, although you don't have to spend much time on Trustpilot to see that sometimes people are precisely as horrible as you suspect. Trustpilot has suffered as much as anyone from the spectre of fake reviews and businesses gaming the system. We'll look at where they come from in Chapter Eight, but suffice to say that Trustpilot's endless assurances on their site that everything is tickety-boo suggests how hard they are having to work to fend off the doubters.

But individual reviews are not the point of Trustpilot (or its competitors Feefo, Reviews.io, Yotpo, Loyalty Lion, Okendo, PowerReviews, BazaarVoice and Rivo). All of them only make any sense in the aggregate.

If I want a car accessories store in London – and today, as it happens, I do, because my windscreen wiper just went kaput – a Trustpilot search instantly serves up businesses with several thousand reviews that average more than 4.5 stars. More than that, I can see that Car Accessories Store 'A' replies to every review, good or bad. This car accessories store is responsible and responsive; they are human, they are flawed, they are trying to do better. And they use star ratings, so they are playing the game.

Amazon's insight that we noted above is relevant here – one-star reviews and the responses they elicit are as revealing about the company, possibly more revealing, than the glowing encomia. It's good to admit and confront your flaws. No one expects perfection, but honesty and a real human being at the keyboard are highly prized.

Ironically, Trustpilot now has a feature which summarises all of this cherished humanity in car accessories ... using AI. 'Reviewers overwhelmingly had a great experience with this [car accessories] company,' a robot assures me. 'Customers appreciate the company's commitment to providing excellent service, with many highlighting the helpfulness and responsiveness of the staff ... Overall, reviewers are very satisfied and recommend the company.'

Obviously, this is anodyne AI bumf that tells you not a lot, but the very fact that AI has been able to pull together a report at all on the merits or otherwise of Car Accessories Store 'A' is the point. The conclusions are founded in volume; they come from bags of stars, numbers, medians, means and weight of data. It's the social proof paradigm we saw in Chapter Two. Collective weight of opinion – founded on many of those opinions being in one place – makes Trustpilot trustworthy. It doesn't really matter what those opinions are.

Trustpilot has the keyword 'trust' in its name, but like most multinational corporations, it is not in business to make friends. In 2021, it was listed on the London Stock Exchange, with an IPO that valued the company at $1.49 billion. Helping his parents avoid hotels with dodgy loos has netted Peter Holten Mühlmann a great, gushing fortune. There is a lesson for all of us sons there – and possibly also for some plumbers.

With all those millions sloshing around, the most obvious question is how Trustpilot makes its money. But the next most obvious question is the more interesting one: does how Trustpilot makes money compromise its trustworthiness? Or to put it another way, can a publicly listed company with statutory responsibilities to shareholders be trusted as your pilot through the stormy skies of consumer choice?

The two answers are connected: Trustpilot charges companies to list on its site. But that means it needs their business and, if being listed on Trustpilot is essentially just going to buy those companies entry to a slanging match, then they'd probably rather not be listed at all. So does Trustpilot and its clients have any control over what ratings appear? Yes, at least they claim they do. If you sign up for Trustpilot as a business, you get a degree of protection, according to their website:

- '**You can flag reviews which breach our guidelines:** Businesses can flag reviews that breach our guidelines (e.g. harmful content, personal information, fake reviews), but misuse of the flagging tool is not tolerated.
- **No fake reviews:** Fake reviews are strictly prohibited and will be removed.
- **Claim your business profile:** Businesses can claim and manage their Trustpilot profile, adding details and responding to reviews.
- **Send review invitations fairly:** Businesses are encouraged to invite customers to leave reviews, but invitations must be fair, neutral, and unbiased, with no incentives offered.
- **Reviewers own their reviews:** Reviews are owned by the reviewers, who can edit or delete them.
- **Respond to Reviews:** Businesses should engage with reviews professionally and politely.

- **We'll take action if you misuse our platform:** Misuse of the platform can lead to actions such as account suspension or profile warnings.'

Reading those guidelines neatly summarises all of the problems Trustpilot, and ratings sites like it, face. How does a business or Trustpilot itself or we the people know when a review is fake? If computers can make deepfake videos of Tom Cruise that look more like a real human being than the actual Tom Cruise (if, indeed, that is him), then surely they can knock up a few lines bigging up or slagging off a product. (Answer: they can. See Chapter Eight.) How can Trustpilot be sure that no incentives are offered for five-star songs of praise? (Answer: they can't. See Chapter Eight.) And what action can Trustpilot take to enforce the line where skulduggery becomes malfeasance and fraud? (Answer: not a lot but they claim they're trying; see etc.)

Trustpilot, as I've mentioned, is certainly not the only game in town. Companies like Feefo, Yotpo, Reviews.io and Sitejabber ('Make Your Voice Heard') offer a similar service, but there are some subtle differences in the trust equation.

Feefo says they 'believe that people should see unbiased and honest reviews on the internet from real customers like you'; there's a big, silent stress on the word 'real'. Their point is that Feefo (slogan: 'stars to help guide you') operates a closed system. Yotpo is the same. They only take reviews from customers who've been invited to leave a review by the business they've just bought from. Feefo and Yotpo say that this means only 'real' customers who've actually made a purchase, rather than some furious keyboard warrior with an axe to grind, gets to have their say about your business. With Feefo and Yotpo, you get to 'move away from a confusing, overwhelming and stressful reviews landscape' as Feefo puts it[15] (even though with their presence they make that landscape more confusing) and free yourself from all of the nitpickers and nutjobs.

This contrasts with Trustpilot and Reviews.io where anyone can have a pop: anyone who has had a 'genuine experience' with a business, leaving aside what a 'genuine experience' means, can leave their feedback. That's more like the Amazon model, where amassing ratings and reviews comes first, no matter where they come from.

Reviews.io operates an open system like Trustpilot, but in addition they incorporate reviews from other sites, making them a sort of mega-aggregator or multi-limbed ratings hoover.

Naturally, all of these sites think their way is the best way. Trustpilot pours scorn on Feefo because it says a closed system is like marking your own homework.* But Trustpilot also snubs its nose at Reviews.io, saying that while an open system (like Trustpilot's) is great, consolidating other sites into your TrustScores is dodgy, because those external sites aren't vetted and can therefore skew the final analysis.

In this, Trustpilot are espousing the doctrine of cake-ism, both having it and eating it. They are saying that their vetting system is superior to a closed system, but that a *too* open system, which is self-vetting, is also a bad thing. Trustpilot think that they, and only they, have the balance set between letting everyone rate to their heart's content and monitoring all those ratings for quality: not too heavy, not too light, just right.

Trustpilot claims to provide 'radical transparency and openness, filtered through an important lens of verification', but they're basically

---

* The UK government has questions for Feefo too: in March 2026 the Competition and Markets Authority (CMA) opened an investigation into Feefo as part of a more general crackdown on fake and misleading reviews. At the time of writing the CMA said it had not reached any conclusions about whether consumer law had been broken.

asking you to trust them on that one. How many of their reviews are bogus? It depends who you ask. Trustpilot will proudly tell you how many millions of fake reviews they have taken down this year, while independent sites like Fakereviewwatch.com will tell you that is just paying lip service to transparency and the whole thing's a bust.

The proof of all this, however, is in the pudding, or the cake, or the porridge: Trustpilot has much more porridge than all the others. It has a far larger market share and global recognition than its competitors* and, as such, has become the pole star in the ratings world. (And in amassing all of this data, Trustpilot has learned a lot about human beings. It is a regular, often entertaining source of useful business data and stupid pub quiz stuff. For example, Trustpilot knows that the most regular complainers, people who leave one-star reviews, are called John.† There are no women's names in the top ten at all. Hard to think why.)

None of this gets us past the essential contradiction behind all commercial ratings sites and, yet again, it comes down to a matter of trust. Can you help consumers and businesses at the same time, or is the nature of commerce always robbing Peter to pay Paul? Can both poacher and gamekeeper trust Trustpilot?

Their response has been to double down on their own trustworthiness. Companies on Trustpilot can't offer incentives or pay to hide any reviews, which is why you can look up Trustpilot on Trustpilot and find out whether the self-described 'universal symbol of trust' can itself be trusted. My favourite out of all of the reviews on the site‡ is the 1.43 stars from 790 reviews awarded to Trustpilot itself, 'indicating

---

* In 2024 Trustpilot's published yearly bookings were $239m, up 23 per cent year on year.

† Also my middle name.

‡ Disclaimer: I haven't read all of them. You shouldn't either.

that most customers are generally dissatisfied with their purchases. Consumers complaining about Trustpilot most frequently mention customer service, bad experience and integrity team problems. Trustpilot ranks 54th among Internet Safety sites.'

In spite of there being 53 safer internet sites than Trustpilot, it makes me trust Trustpilot that little bit *more* knowing that they are still prepared to post reviews saying you shouldn't trust them. As I tell my children, being honest about your dishonesty is always the best policy. It also points to what Trustpilot is all about, which is, as the name suggests, the struggle for honesty and credibility in a world full of fakery.

Like so much of the internet, the idea behind Trustpilot is sound, but 20 years of exposure to human greed and the machinations of late capitalism have contaminated the waters.

The site claims that it was founded on a simple yet powerful idea that is more relevant today than ever: 'to be the universal symbol of trust, bringing consumers and businesses together through reviews'.

This, when you think about it, is quite the claim. A universal symbol of trust is a very powerful idea indeed, but it is not a simple one. Throughout human history, there has never been a *universal* symbol of trust – different cultures have come up with different symbols to signify credibility and certitude. In the UK, the three letters NHS are trusted more than any elected government has ever been. In the US, Dawn Dish Soap tops the list of brands trusted to 'do the right thing', some way ahead of all the brands whose watchwords are you can trust us (although it only gets a 59.8 per cent rating, which isn't the most squeaky-clean endorsement).[16]

In the business world, trust merits loyalty and thus commands a premium. But it's vital in the world at large too. The Scottish philosopher John Locke thought that trust was central to consensual democracy. Trust, in Locke's view, was dependent on God. If this trust was breached, then men had the right to rebel against their sovereign. Locke's universal symbol of trust, therefore, would have been a cross – both the cross of Jesus and the cross in the voting box.

All of that may seem a little lofty in the context of deciding which vacuum cleaner or windscreen wiper to buy, but in a post-religious society, it follows that if you take away God as the higher power overseeing people's obligations to one another, and if you then take away people's belief in government, or John Lewis, or whatever else you used to trust in, then you need something to replace him/her/it.

In the UK, royal warrants were for years a solid-gold star rating bequeathed by queens and kings. These days, you can get a royal warrant as a 'supplier of recorded telephone messages', which rather undermines the whole premise. There are more than 800 royal warrant holders; King Charles can't have tried them all. And, anyway, given some of his own life choices, he might not be the watertight arbiter a sovereign used to be.[*,†]

---

* A royal warrant is initially granted for up to five years to a named individual at a company, known as the 'grantee'. There is a Royal Household Warrants Committee who say they review all royal warrants in the year before they are due to expire. But there is no list published detailing *when* any of the current warrants are due to expire, meaning it's impossible to know if anyone is being assessed. The committee is drawn from the Royal Warrant Holders Association, which first formed on 25 May 1840 when a gathering of 'Her Majesty's tradesmen' held a celebration in honour of Queen Victoria's birthday. They later decided to make this an annual event. Anyone looking to get themselves a royal warrant, be they cheesemaker or pest control officer, should go to www.royalwarrant.org.

† We know, for example, that King Charles has a problem with pens. In September 2022, while signing a visitor's book on an official visit in Northern Ireland, he got very irate about a leaking pen, saying, 'I can't bear this bloody thing.' Because we do not know the brand of pen he was using, we can't say if that penmaker has (or had) a royal warrant.

But if you don't trust in the king or a deity, then who do you trust? It's often remarked that shopping malls are our new cathedrals, but less often that branding and marketing themselves are acts of faith. You have to believe at one level that the white cotton T-shirt with the little crocodile on it is somehow worth 100 times more than the white cotton T-shirt from Primark with no crocodile on it.

You can go back to reciprocal altruism and game theory, and conclude that 'natural selection crafted an ever-expanding web of affection, obligation and trust out of ruthless genetic self-interest', as the sociobiologist Robert Wright says in *The Moral Animal.*[17] In other words, that trust is Darwinian and is founded somewhere deep in our selfish genes.

But the fact remains that, in individual scenarios, people don't feel the non-zero-sumness of the prisoner's dilemma at work. They just wonder if that website, doctor or Vinted seller is legit. Take away God, or deference, or authority figures, or even respect for leaders and politicians or class boundaries, and you are left with a credence void. It breeds unease, when all the human mind really wants is security.

Those of you who have ever attended Catholic Mass might recognise what's known as the Credence Table – a literal table where the items for the Eucharist are all placed side by side. The origins of the Credence Table go back to the Middle Ages. It came from the Medieval practice of 'tasting' food and drink for poison, with the table serving as the location for an official to sample the offerings before the meal and hopefully not keel over and die.

Trustpilot and its ilk are our modern-day credence tables, although they earn their credence not from divine consecration but from stats. 'Ten thousand users can't be wrong' is a sales pitch as old as time and that's because it taps into the doctrine of social proof and collective wisdom that we looked at in Chapter Two. It is no longer enough to just be roundly well-known as a good company that makes good products. You need numerical proof – x number of stars awarded z thousand times.

In the UK, for years the most trusted retailer was John Lewis, a haven for the middle-classes and renowned for quality, value, knowledgeable staff and good warranties; they basically represented the great British sense of fair play in retail form. Naturally, none of this stood a chance in the bear pit of e-commerce where price, delivery time and returns policy soon swamped any other consideration. People would talk about how much they loved John Lewis before nipping over to Amazon because their Prime subscription got them free next-day delivery. Unsurprisingly, John Lewis started using star ratings in 2011.*

Similarly, magazines like *Which?*, *Good Housekeeping* and *Consumer Reports* got by for years trading on what used to be called expertise. They offered good advice and you knew it was good advice because, well, *Which?* and *Consumer Reports* had a good reputation. Everyone knew it. Their credo was simple, as summarised by a line that *Consumer Reports* uses to this day: 'Making Buying, Owning and Living Better for Us All. Dedicated Experts. No Corporate Influence. Here to Fight for You.'

Who else is on your side these days? The Better Business Bureau in the US says it is. Its mission is to be 'the leader in advancing marketplace trust', which they do by listing ratings, reviews and complaints that they guarantee are independently verified.

Claire Rosenzweig is the president and CEO of Better Business Bureau Serving Metropolitan New York. She told me: 'It's just really important for consumers and businesses to know that when they

---

* As a thank you, or perhaps a bribe, every review you write is entered into a monthly prize draw for the chance to win a £100 John Lewis gift card. Sellers need your ratings and reviews.

are coming to a BBB review, not only do we verify and make sure the customer is real, but we go back to the business and say, "Do you know this interaction?" We just want to make sure that the review is real, that the response is authentic and that consumers can trust what they're reading.'

But while *Which?*, *Good Housekeeping*, *Consumer Reports* and the BBB are still in business, and while no one doubts their expertise and credentials,* they have been superseded by the preponderance of consumer reviews and ratings sites. These claim no expertise other than the wisdom of the masses. They are fuelled by the modern belief in the power of data. And in the internet age, this has proved more compelling than the old notion of paternalistic trust based on experience and reputation that says you can trust us because we are known to be trustworthy.

The numbers game is so powerful that, these days, no advert is complete without a Trustpilot, Feefo, reviews.io, Yotpo or similar tag. Their star ratings have become as much a part of marketing as advertising, catchy slogans and celebrity endorsement.

The ubiquity of these cumulative star ratings reflects a broader historic and cultural change over the last decade. In the pandemic, companies like Trustpilot had what you might call a good war. As global lockdowns drove both commerce and services online, there was more need to weigh up who you were buying from (and many more online options to weigh up). At the same time, more people were stuck at home on their laptops with hours stretching out before them to do some reviewing, both reading and writing.

---

* Actually, several people do, with the BBB in particular coming in for criticism for being in hock to paying businesses ahead of consumers according to some people I spoke to.

The rise and rise of Trustpilot and its competitors also answers a question. If e-commerce, spurred on by everyone being confined to their bedrooms during the pandemic, has turned the world into what marketing people like to call 'the infinite shelf', with more choice coming from faceless retailers and just appearing on your phone or laptop from who knows where, how do you orient yourself in this dizzying carousel? Retail has come to mean not just the shopfronts you recognise on the high street or not even the well-known online brands (that also advertise on the high street). Retail can now be a strange non-word like Temu or SHEIN or AliExpress, popping up on your phone ads with the strangest products, like robot potatoes playing flutes wearing inflatable glasses. When products you didn't know you wanted appear at prices so low, do you really need to grab them before they're gone? When it looks too good to be true, is it?

Star ratings on sites like Trustpilot purport to provide some sort of cognitive anchor. They're best used like a form of ID card for when you have no idea who this vendor is and want some back-up. It follows that the more businesses they have signed up, the more useful they are.

If the numbers game is what gives the collective ratings of sites like Trustpilot force, it follows that the best review aggregator should be the biggest one. And that leads us to Google.

★★

No one commands more global eyeballs than Google. More than half of *all* online reviews are on Google.[18] The difference between Google and other review sites like Trustpilot is that anyone can post a review on Google of anything. Trustpilot reviews are only offered to confirmed customers, most of whom have to pay Trustpilot to feature. Google is, by some way, the most popular search engine and therefore most people's portal to the internet (and hence the outside world).

Google publishes star ratings and reviews alongside its search results, which makes it a mind-bogglingly powerful influencer. Eighty-eight per cent of us read Google reviews to determine the status and quality of a local business before engaging with them,[19] while 84 per cent of us trust those online reviews as much as any personal recommendations.[20] Google is the world's largest metaphorical credit facility.

If you google anything (and very few companies in human history have become their own verb so quickly), someone will have written a Google review of it, with stars at the top – and generally a withering assessment based on a disappointing interaction beneath. I'm doing it now, with the first thing that comes into my head, which, because it's nearly lunchtime, is a chip butty.*

Googling 'British chip butty' yields Kate Hackworthy's recipe at veggiedesserts.com ('It's carb heaven and you NEED to give it a try'), which has amassed 15 reviews,† every one of which is a straight five stars.[21]

Reviewer Margaret says, 'A classic! But for me, I have [sic] to make your own chips in a chip pan that contains lard. The only way to make them.' This, given Kate doesn't make them in lard, and given this is veggiedesserts.com, should surely mean that Margaret withholds a star or two. But she doesn't.

Likewise 'bill marsano', with his low-capped insouciance, 'recommends tallow', another triglyceride meat duvet, yet still thinks Kate's version in lily-livered vegetable oil is worth five stars.

---

* In the UK, a 'chip butty' is a sandwich filled with chips. And in the UK, 'chips' mean thick-cut fries. The butty originated in fish and chip shops in Britain in the 19th century, though the exact origin is (hotly) disputed among carb aficionados.

† These reviews are on Kate's site but are also viewable on Google, as is often the case.

In fact, everyone who cares to comment on veggiedesserts.com about the heavenly chip butty gives Kate's version a five-star Google review, which does make you wonder if Kate Hackworthy is filtering her reviews or if people realise that other star ratings are available.

Kate Hackworthy (and I'm not picking on her, I just like chip butties) is herself rated, one of Vuelio's top ten blogs in 2018 and 2019 with a commendation from the Food Magazine Reader Awards and, yes, a 4.7 TrustScore on Trustpilot. I'm not able to find if Kate has filed any reviews herself, but all things happen for a reason and that reason is normally Google. Google's search algorithm has most likely placed Kate's chip butty recipe at the top of my search list precisely because she is well-rated herself. Stars are circular, a self-perpetuating zodiac that goes around and around the mountain like the Paramount logo (more on this in Chapter Seven).

Google reviews, like star ratings, were initially for travellers; they originated out of Google Maps. The idea, just as it was on the Home Tour and for Mariana Starke and Michelin in the 1930s, was that if you were going somewhere, you'd want direction – not just in literal pointers but in qualitative terms too. As Google Maps became the pre-eminent source of route-finding, it was an obvious next step for it to offer recommendations as much as navigation, just like the *Michelin Guide*. As such, when Google Local was launched in 2004 before merging with Google Maps, business listings were aggregated from other review sites like Yelp, Tripadvisor and Zagat.

In 2007, also the year that Trustpilot launched, Google began allowing users to leave their own reviews directly on business listings. At the same time, the slow, inevitable tech seepage of a website going from useful-to-the-consumer to lucrative-for-the-shareholders began, as Google Places was introduced as a standalone platform for businesses to manage listings. Businesses could now respond to reviews, while users could leave one-to-five star ratings.

In 2013, in another page straight from the big tech playbook, Google simply bought Zagat and swallowed its many years of restaurant experience into Maps. Zagat used to use a (much-loved) 30-point system to grade restaurants, but when Google tried that on the mass market, it was deemed too confusing and quickly dropped in favour of good old one-to-five stars.

By 2014, the wholesale absorption of star ratings into a facet of every business's profile was just about done. Google Places was rebranded as Google My Business, streamlining the way businesses managed listings, photos, opening hours and reviews. Reviews became more prominent in search results and – crucially – on Maps.

And so the Google review ecosystem – which, because of Google's towering scale, is now the de facto review ecosystem for the entire world – was set. You'll know the parameters all too well: whether you search for it in Maps, Search or via AI, pretty much any business, service or place gets a star rating straight after its name. Written comments, photos from users, upvotes or likes on helpful reviews follow (though, of course, Google's algorithm decides 'what's relevant'), alongside business responses, flags for inappropriate reviews and, in the last few years, AI summaries of all of the above. This is the framework and first portal for almost all human life online. But that star rating comes first.

The Google review process is open to all. I like to review as much as possible, including a motorway service station I visited yesterday on the way back from a theme park. (I also reviewed the theme park.) If you're a Google reviewer too (and many millions of us are) you'll know that when you post a review you get a pat on the back, in the form of a Google graphic that appears in a dialogue box and fires off some party poppers, with the words:

'You have earned +11 points for helping others.'

Click through to your profile page on Google Maps and if you've been rating zealously you will discover that 1) Yes, you have a profile on Google Maps; they're rating you and 2) You are now a 'Local Guide Level-something'.

It feels great to be important, doesn't it? Further exploration reveals that 'as a Local Guide, you earn points when you share reviews, photos and knowledge on Google Maps. Those points lead to higher levels of the program, as well as benefits like early access to Google features and invites to special events. At Level 4, you also unlock your first Local Guides badge, which helps your contributions to places get noticed.'

And helps you get noticed. Everyone wants to feel seen.

Why does Google bother with all this flummery? They say it's because they are 'dedicated to helping you discover the world around you', which is nice of them.[22] But in truth, just as with Amazon as we saw earlier in the chapter, by leaving a rating and review, you're working for one of the world's largest companies for free. You're spending your time making their maps app more useful, while providing Google with more and still more data about its users that it can sell on to advertisers. Google, it's always worth remembering, makes more than 85 per cent of its $337 billion revenues from advertising.[23] Those things you willingly tell it about yourself when you submit a rating and a review – what you like, what you don't, where you've been, what you've bought – are incredibly valuable. In addition, the rise of AI means that content that you submit, including reactions and related metadata, may be used to provide, develop and improve Google products, including machine-learning technologies and generative AI.

Google's calculation is that users like getting stupid points and being called important and upgrading from a Level 3 to a Level 4 so much that they will just hand all that data over.

And we do, all the time, all over the place. Google Reviews take in government institutions and public services, places of interest and tourist attractions, home-based businesses (I live two doors down from a T-shirt seller, apparently), cash points, my local café, your local café and basically anything else you can think of that someone, somewhere, might need at some point. Here, then, is the rated society.

We've talked about how that rated society is a self-sustaining hall of mirrors, so my next port of call is to see if Google allows reviews of Google. And it does! Search Googleplex at 1600 Amphitheatre Parkway and you'll be delighted to find that the top listing at the time of writing, courtesy of Local Guide 'That Bearded Mechanic' tells you in no uncertain terms that 'Google is ABSOLUTELY useless when it comes to them following their own policies regarding fraudulent and spam reviews on business pages. They do absolutely nothing to follow through and remove them when reported. Even when escalated with proof of the false review based solely on spam and to attack a business. Do better.'

Google Reviews covers so much with so much of this scrupulous transparency that it is more interesting to look at what it doesn't cover. In the UK, you won't get a star rating for schools, for example. This is because, in February 2025, Google agreed to remove all user comments and ratings of UK and Irish schools from Google Maps after a petition organised by an exasperated Merseyside teacher called Justin Cowley. Cowley was fed up with his school being judged unfairly on Google Reviews. Parents, he said, should be able to make 'an informed choice, not an outdated choice based on historic data'.[24]

Anyone who ever looked at a school's Google rating in the past will know that there was a bit of playground tomfoolery going on here too: school reviews had become repositories for the best offerings from all those online cheeky chucklers looking for an audience, many of whom had been at (or claimed to have been at) the school and were now grabbing the chance to bite back.

Eton College, for example, one of Britain's most prestigious schools, had garnered a 3.7 star rating for its annual £63,000 gilt-edged education. As ever, there was wry wagdom to be enjoyed: 'I expected more of the gardens,' read one review.

In an email sent to schools, Google said: 'This is a change for general education school listings in the UK and Ireland to prevent unhelpful or prank reviews ... Existing reviews or ratings for your school will be removed, and users will not be able to submit new reviews or ratings.'[25]

'Unhelpful' was doing a lot of heavy lifting there. Some reviews referenced teachers by name; moreover, parents were using Google Reviews to band schools and education with any other consumable.

The question was whether allowing people to rate and review schools in the same way they rated sausage rolls was either 'helpful' or 'fair'. Google had to concede that it was probably neither.*

In our survey of the rated society, this is an inflection point. It shows there is a sketchy perimeter to the rating of everything that society has deemed should not be crossed. Workplaces are fine. Glassdoor.com is a Trustpilot for work, with an annual 'Best Places to Work' list that companies strive to be on (in spite of the fact that it lumps Crew Car Wash, tech company Nvidia and the Church of Jesus Christ of Latter-Day Saints all together as potential employers to consider). Tripadvisor rates travel, which as it encompasses people and places, is close to the edge, albeit in the sphere in which star rating first began.

The trouble comes when it gets personal. What happens when people start rating other people, based on the sort of people they are? And that's where we're headed next.

★★★

* It hardly needs saying that there are still websites available that allow people to rate and review schools. Locrating, for example, is a British 'school finder and catchment area checker' that allows verified users to leave star ratings and reviews, just like Google used to.

Chapter Five

Rate Me

I have been a 'member' – or, more realistically, a healthy revenue source – for eBay since April 2002. You're probably one too. In that time, I have accrued 167 positive feedback ratings and this, I am proud to report, has earned me a fetching light-blue star by my username. In the quick-fire reviews left by the people who have bought my clutter, I am variously described beneath my star rating as 'pleasant', a 'great communicator' and 'an absolute pleasure to do business with'.

I'll admit, that's great to hear. These are commendations of character as much as commercial nous. Basically, I'm one heck of a guy and it's my promptness, volubility ('great comms!!!') and candour ('item was exactly as described') that prove it. No one talks about price; no one says, 'grabbed myself a bargain off of this idiot. Pile in, people.'

Yet eBay stands as an example of where star ratings can get personal. The point of a star rating is to provide an overall objective measure that other people can use for assessment. Or that was supposed to be the point. But as stars have become the grading currency of the online world, they have become a measure of morality. There are now many examples whereby whatever you seem to be rating and reviewing – that toaster you just bought on eBay; that dog-eared paperback on Amazon; that last trip on Uber – what you are really rating is a human being. What your one-star put down is saying is, 'I think so-and-so is a dick.'

Here lies a classic example of tech titans not quite understanding the consequences of the disruption they were so cheerily unleashing. When Jeff Bezos channelled Jean-Paul Sartre in his comments about how we are defined by our choices, with customer reviews being the pointers for direction, he perhaps forgot that there's an overlap between humans as social animals and humans as consumers. We judge people in the same way that we judge products. But people are not products – or they shouldn't be.

You'll have encountered the problem with businesses like Uber or Discogs or Airbnb; anywhere that requires interaction, credibility, praise or provenance has wound up with a star rating system as its currency.

There are myriad stories of how star ratings have got personal. You'll probably have a few of your own. When I first bought something on Discogs, where I tend to my vinyl addiction, I was labelled 'joker' (not in a good way) and given a one-star thumbs-down because the record I'd bought had sat at my post office and been returned to sender. I had then left the seller a 'neutral' review because the record never arrived. Big mistake. That early low rating rendered me unable to buy any more records, because vendors can filter out buyers with a bad rep. A phone call – as opposed to an online star rating – could easily have sorted this out. Instead, Vinyl Queen N14 now really hated me and I wasn't that enamoured of them either, given that their condemnation was public and had effectively blackballed me from the world's largest record-buying community. All this was basically because they thought I had rated them unfairly. Which I had, it turned out.

On Airbnb, there are countless examples of how professional criticisms can descend into personal slights, all aired in public (which is why I can quote this frighteningly commonplace horror show from Tabby242 in Tucson):

'I had an awful guest experience,' Tabby242 wrote, 'and tried to settle it with the guest personally. She stayed at a reduced rate and then needed to extend her stay. We gave her an even bigger rate discount. When she left, she left the place a mess. Dirty, greasy dishes, pots and pans. All the furniture was in disarray, both sets of blankets/quilts had spots and stains on them, there were crumbs in all the beds on the mattress pads (which had to be hand-picked off), and she ruined a set of sheets.'

Tabby242 asked for $50 to replace the sheets. Uh-oh. The now-roused-to-anger guest 'did not understand why she should have to pay for her sheets because I was a woman and the damaged sheets were a result of a "woman thing".'

You can probably see where all of this is leading, and also that it is not going to end well.

'After much internal debate, I decided to warn other hosts and gave her an appropriate but still not brutal review. She then gave me a bad review, lying about my place and me. Prior to this review, I had 5-star reviews. Now people think my place is noisy and I am rude. My bookings have almost stopped dead in their tracks after her review.'

Consistent with the never-ending nature of both the internet and the ratings-go-round, this wasn't the end of it. Tabby242's travails received a slew of responses on the Airbnb Community Centre. Top Answer came from John915, a level-four host in New York:

> 'Some guests are just horrible and/or petty (and I'm sure a handful of hosts are not so great either) but the best thing to do is learn from it. My advice:
>
> 1) Do not post a review first – Regardless of what you do, you sometimes get less than 5-stars! If you've had a less than fantastic dialogue with a guest, posting a review prompts them

> to do the same – in your case, she knew you were upset about the sheets & was probably defensive about being asked to pay $50 for an accident – If she thought you were upset with her as a host, I am not surprised that she didn't give you a fantastic, glowing review in response to the notification that you had left her a review.
>
> 2) Don't sweat the small stuff.
>
> 3) Just buy a $10 set of sheets from Wal-Mart and consider it a "cost of doing business" if they are ruined at some point.'

Needless to say some other happy giggler then instantly posted '$10 sheets? HAHAHA.'

Unusually for online comments, John 915's Top Answer was actually quite sensible, at least the first bit. John was saying your best bet is not to review at all, because it always ends up getting personal. By saying anything about someone else, you are casting judgement – and people don't like to be judged.

The problem is that it's quite hard not to rate with abandon. The digital economy thrives on these judgements, because it needs all this feedback for data mining, market research and just to keep the cogs of commerce turning as we've seen. The danger zone is where the personal and the professional overlap. Being a 'good' doctor, vet, nanny, teacher or binman is not merely a measure of your aptitude or ability. It is a measure also of whether people like your manner. This is why businesses don't just employ staff on the basis of their test scores – they want to interview them first. Countless (depressing) studies have proved that good-looking people with a fast tongue and an affable air are assumed to be more capable, not least in politics.[1]

At some point, however, the qualitative – are you nice? – runs up against the quantifiable – are you productive, successful? – and that clash feels callous. This is why services like Rate My Teacher, Rate My Professors, RateMDs, Doctify and the like frequently prove controversial.

Housekeep.com, for example, is a UK-based app that provides domestic cleaners and maids for messy apartments. The way this used to be done was via small ads in newsagent windows, or more often by word of mouth and a shared phone number. Housekeep says it offers a guarantee of timely, quality work backed up, of course, by consumer feedback. It also promises that if your preferred cleaner can't attend, then it will simply slot in someone else from its books.

You can see how this could easily run into trouble. A housekeeper is an individual, not a product substitution. They are generally a person on their own, reliant on piecemeal work and for whom a single one-star rating (and we know that star ratings tend to gather round the extremes) could harm future bookings. Yet that rating itself can be delivered by an irrational actor or an irate old crank with ridiculous standards who just doesn't like the person who's been sent into their house on their own for what can be a quite personal interaction. The 'Rate Your Clean' button that's emailed to Housekeep users after every mop and iron gives inordinate power to one individual over another. Power wielded without accountability tends to lead to abuse.

I spoke to a number of Housekeep cleaners about how ratings affected them (they all wished to remain anonymous in order that they didn't attract another poor rating).

'You work for so many different people,' one said, 'and sometimes they just don't like you, or they have different standards for what a cleaner should be doing, or they are just in a bad mood. But especially when you start on the site, a bad review can ruin everything.'

Housekeep summarises your thoughts in a single word that it generates beneath your star rating, and three stars – which to many of us means 'absolutely fine', indeed numerically above average – is 'disappointing' in Housekeep's evaluation. It's certainly disappointing if you are one of its cleaners who take home just around minimum wage, while Housekeep charges a further 44 per cent mark-up.

A similar practice is commonplace with telephone helplines. You will have been subject to the endless entreaties for ratings after every call to a customer service line (which, you can't help but notice, may be recorded). To take one example, workers at the Carphone Warehouse in the UK used to be rated after every call – and would miss out on bonuses if they failed to get enough high marks.* How hard do you think about the numbers you're doling out in these ratings, though? You're probably not aware that a low mark given because, well, you were tired, or you couldn't understand their accent, or just because, may lead to someone being paid less.

Vets, like doctors, stand or fall on personal interaction. A large part of the job is a form of bedside manner, soothing emotional owners as they come to terms with the fact their pooch is toast. Part of that caring, sharing manner these days is often delivered via feedback to feedback on Google Reviews, which can make or break a fledging business reliant on personability. So when the following review came into a small practice in the UK that was used to wall-to-wall five-star ratings, it caused some consternation.

'Appalling!!! 😠. WE would definitely NOT recommend this place!!! They don't deserve 1 star [although the user gave one star], we should have thrown our money away 💸 💸 💸 🗑 than waste it on this vet practice!!! Avoid.'

---

* This is no longer the case. A Carphone Warehouse telephone salesperson I spoke to said that good ratings were noted in end-of-year appraisals but not directly linked to bonuses. The company now encourages callers to leave feedback and a rating on ... Trustpilot.

The reviewer was a Google Local Guide Level 5, with 1,179 points to their name, and so their rating had to be treated with the respect the vets might themselves have hoped for. As such, the vets replied promptly with a conciliatory response asking the client to get in touch directly. Inevitably, they didn't.

Yet, on further investigation, this reviewer's reviews always carried a certain tenor (as well as the same star rating). The local transport museum got one star ('Maybe the cars/vehicles are fascinating to some people, but the place smells weird.') They had similar views on McDonald's ('Awful experience, food portions are never worth the cost they charge in my opinion and even the entrance doorway still broken, at this location'), shopping centres and virtually every store they had ever been in. In fact, pretty much every time they left the house and committed the experience to Google, it seemed to be a hellish ordeal.

But particular bile was reserved for vets. Several local veterinarians had a whole world of potential future customers told in the strongest terms to 'Avoid, Avoid, Avoid!'

It's easy to cast off these types of hissy fits as howling at the moon, but when rating small companies based on one-to-one interactions, they can come very close to being personal slights. I got in contact with the practice manager of the vets that had been told they didn't deserve that single star.

'We are a small practice with very caring vets who take a great deal of pride in the service we provide to our clients and their pets. The client in question had shown no dissatisfaction at any point during their visit to us. In fact, we had advised them against spending unnecessary money or risking their pet under anaesthetic as the operation they were investigating was not in their animal's best interest.'

Understandably, this vet's practice was shocked to see the one-star rating pop up on Google initially without any comments explaining why it had been given.

'As we would with any feedback, we investigated this. Internally, there were no indications as to why this client would be dissatisfied with their service and the vet who had seen them was upset to think that they could have somehow caused offence without meaning to and stressed to what this might make us think of them too. We responded to the rating, expressing our upset at hearing their level of dissatisfaction and asking if they would like to call the practice to discuss what had led to them leaving us only one star. Instead, they responded with a barrage of online feedback that you view above. We delved deeper and realised that they were a serial offender with regards to expressing strong negative feedback to businesses online.'

To which the world says, 'whatever'. As always with Google Reviews, the vets can't remove the slur on their business and they also can't respond with their version of events. This is partly due to client confidentiality (which clearly does not work both ways) and partly due to not wishing to enter an online 'he said, she said' where everyone gets angrier and angrier and which achieves nothing.

Increasingly, though, small businesses are reliant on reviews to exist at all. Danielle Dybiec is the founder of Nine Muses, a travel agency based in New York City that is, she freely admits, a one-woman band.

'Our sole employee ... you're looking at her,' she says. 'I don't have a storefront, I work from home. There's no way to know that that photo on my website is really me. [It is.] So reviews are a way to prove that I have actual clients who do actual travel.'

Dybiec uses a service called Virtuoso where the reviews are verified by people who have genuinely been on one of her trips. 'Whenever a client comes back from a trip, I ask for reviews. But I let them know I'm a small business owner and that this is really impactful for me. When I get the email from Virtuoso saying the review has been posted, I can't edit them so I let clients know – although it's kind of unspoken – that if you don't have anything nice to say, maybe don't!'

All of this brinkmanship with the anonymous disgruntled takes time – time when the vets could be doing more vetting and the travel agents more travelling.

'Businesses are busy,' says Melanie McGovern, director of public relations at the Better Business Bureau in the US. 'You're running your business. You don't have time to do online reputation management, post to social, you know, comb the internet for any kind of negative review about yourself.'

Vets and travel agents and any agency or worker who is customer-facing in a small business are like doctors and teachers, in that their manner and personality is part of their professional offering. A star rating is therefore inevitably personal.

★

Posting online reviews is also public commentary and therefore potentially libellous. In 2021, a man called Philip Waymouth decided to hire the London law firm Summerfield Browne for some online legal advice. He wasn't happy with their service and so he went on to Trustpilot and left a review accusing the firm of being 'another scam solicitor'.

I know this because it's in the court documents: Summerfield Browne promptly took legal action, stating that the review was untrue and defamatory. The number of business enquiries it received had dropped since the publication of the review, they said. Waymouth hadn't engaged with Summerfield Browne's own complaints procedure before heading on to Trustpilot to vent – although he had offered, possibly foolishly, to remove the review in exchange for a refund of the firm's £200 fee.

The review ended up costing Waymouth £25,000 in damages, payable to the solicitors who had charged him £200 in the first place. Slagging someone off online to make yourself feel better has consequences – particularly if they're lawyers.[2]

This means that discretion is advised when it comes to avvo.com – a US site that offers ratings, reviews and disciplinary records for lawyers across America. A cursory tour of Avvo reveals that there are very few negative reviews of lawyers, possibly for the reasons above: lawyers are pretty handy when it comes to suing.

This does render avvo.com slightly pointless, however; a blanket of positive ratings and reviews helps no one. I wonder how many negative reviews have ever been posted there and then taken down at a lawyer's request. Avvo has stringent policies for what can be posted and, to lawyers, stringent policies are catnip for flagging up the teensiest violation. It makes it very easy to get just about any review taken down.

Even if you don't have a paid Avvo plan, there are external services for removing Avvo reviews.[3] These are charged on a 'per removal' basis, essentially skipping the need to even pay Avvo a monthly subscription of $600 a month in order to net the same results. That's great for the lawyers who want only great ratings, but where does this leave potential clients? None the wiser.

We saw in the last chapter how schools in the UK have managed to extract themselves from Google Reviews, but Rate My Professors, a service for lecturers and university teachers, remains online and prospering.

Rate My Professors is the largest online site for students to run the rule over lecturers, colleges and universities across the US. (It also covers the UK and Canada.) 'Our mission is to provide a safe forum to share classroom experiences to help fellow students make critical education choices,' says their website.

Rate My Professors boasts that it hosts reviews of 1.7 million professors from some 7,500 colleges and universities, and that 'more than 4 million college students' use it each month. Somewhat weirdly, Rate My Professors also acknowledges that the reviews it hosts may not be valid or reliable: 'Considering the reality of our digital age ... we're unable to fully confirm our users' identities, the truthfulness of their contributions, or their student or class enrolment status.' Because Rate My Professors does not authenticate reviewers' identities, there is simply no way to know whether reviewers have actually taken the courses or even met the professors they are reviewing.

RMP, as it calls itself, offers one-to-five-star anonymous ratings from students, ranging from the disgruntled to the merely gruntled to the totally delighted. You don't really need to look at it to guess how things play out on the site. It's the usual mixture of sour grapes, self-selection bias, axes being ground and share-this drollery (Isaac Newton gets a few reviews in the mathematics department at the University of Cambridge, landing the father of modern mathematics a three-star rating and the comment 'bro literally just teaches his own textbook'), but very little that you would call useful.

But it can be harmful. Reviewing RMP in 2022, the professor David Varel wrote that it was 'more quantitative in nature, less substantive, and more clearly intended for administrators to justify actions towards faculty members regarding retention, promotion, and tenure'. This model, he wrote, 'fails instructors, and it fails our students'.

When composing a review on RMP, students rate the course's difficulty, how likely they would be to retake the course and the professor as both an instructor and a person. As Varel noted, the site encourages students' subjective responses and positions students to 'view themselves as consumers assessing the "product" they received', a process that is 'eerily similar to surveys sent out to customers of major corporations'.[4]

This is how star ratings have evolved: people are graded like products.

For example, until 2018, RMP also used a 'hotness' ratings in the form of a chili. They explained in a tweet that 'the chili pepper rating is meant to reflect a dynamic/exciting teaching style'. But they knew as well as their students what 'hotness' meant. (RMP even published lists of the 'hottest professors' across the US every academic year between 2009 and 2017.)

The chili was removed after one of the more predictable outcries in the history of outcries from female professors who said that the rating was sexist. Obviously, there is no connection between ability as a teacher and devastating good looks, but the chili stood as a reminder that when ratings are assigned to individuals by individuals, it always gets personal. And at that point, the ratings cease to have any value as an objective measure. A lengthy study by the Academic Integrity Group of all RMP data, for example, concluded that 'at best, Rate My Professors is nothing more than a popularity contest and "easiness" rating. At worst, it's a tool of discrimination against women and minorities. Either way, it's nothing more than a collection of clickbait nonsense.'[5]

The problem is that people really like clickbait nonsense (the clue is in the word 'bait') and so rating individuals has spread right through the professions. When it comes to professionals like doctors and surgeons – people who are going to literally try to fix you – you have an unenviable scenario where something nonsensical is being used to measure something that is vitally important.

Healthcare systems differ worldwide and don't exist at all in some countries, but systems like that of the US are particularly susceptible to public rating and review. Physicians with a private practice there rely heavily on reviews because they're responsible for their own marketing and advertising. For laser eye surgery, cosmetic surgery, orthopaedics or anything involving obs and gynae, reviews are huge.*

As a result, there are a lot of niche specific websites like RateMDs, Vitals.com, Review Dentists (UK) and RealSelf (for plastic surgeons, whose job, ironically, is to make people look better, but who are now getting their own reputations hacked to pieces online).

As it happens, RateMDs was founded by John Swapceinski – the same man who set up Rate My Professors with its controversial chilis. (He sold it in 2005.)

'It's really easy to find [internet] ratings for books or for iPods or even toasters,' said Swapceinski at the time, 'but when it comes to medical care there was pretty much nothing on the internet. It was definitely a need that was waiting to be filled.'[6]

RateMDs, which is based in California, allows users in the US and Canada to rank physicians like they would those iPods or even toasters. Their slogan, inevitably, is 'Doctors you can trust'. They ask patients to grade their quacks based on three factors – punctuality, helpfulness and knowledge – and then post comments. The site owners then use the information to assign a physician rating (out of a maximum total of five) that it publishes online. These, as you'd guess, run the gamut from the dubiously flattering to the extraordinarily irate.

You can imagine the outcome. Patients loved it and doctors hated it.

'Most doctors don't like the site,' said Swapceinski in the site's early years. 'We get threatened with lawsuits on pretty much a weekly basis.'[7]

The panacea of consumer choice, coupled with an entrepreneurial tech bro's boundless ingenuity, had opened the door to the entire

---

* The UK, by contrast, has a National Health Service that is free at the point of care for all. Because it's free, you can't be too picky about which doctor you get. The UK also has a lot of private healthcare, however: see below.

medical profession having their competence questioned by people who had no medical training themselves.

Doctors are a good example of how the very idea of rating individuals – and what you're actually rating when you do so – is contentious. Patient satisfaction is based on a whole series of metrics, of which patient outcomes are just one. You might have hated your surgeon, but that doesn't mean he or she didn't save your life. Equally, a part of good doctoring is telling people things they don't want to hear. It may take years for the wisdom of the good doctor's prescription to become apparent; it only takes seconds for the patient to express their displeasure online.

At the same time, however, the marketplace in medicine has discovered what the marketplace in e-commerce has known for years: that in order to make sales, you need to have reviews. The system is therefore open to abuse.

Curtis Boyd was a nursing student in California when he first stumbled upon this particular bomb crater. 'I was making $12 an hour at a local hospital, but I was kind of techie, you know – I've been playing computer games my whole life,' he told me. 'There was this cosmetic surgery doctor on the ER who started complaining about a fake review that was hurting his private practice. He wanted it taken down.'

The doctor who had suffered the scathing review had tried hiring lawyers. He'd hired reputation management. No one could help him.

'I told him that I might be able to remove it for him. He jokingly said he'd pay off my student loans if I did. I told him how much in student loans I had and he said he'd probably lost more than that already in the week that the review had been up, since it lost him seven or eight new consultations.'

The next day, Boyd called the surgeon's office and asked him if he was serious about his offer he made in the ER. He said he was. 'I spent the next 72 hours in research mode, flagging and disputing the bad review like crazy. I couldn't get it off. The review administrators were adamant, and I was really discouraged.'

So Boyd flew to San Francisco and parked in front of the door of the reviews site, buttonholing the staff as they went in.

'I solicited hundreds of people as they walked in and out of that building and I asked them, "Do you work here? Do you know how to remove a fake review on your site for a doctor in LA?" And they're like, "Are you hungry? Are you homeless? Are you crazy?"'

But after two days of picketing staff in San Francisco, a staffer at the ratings site finally took pity on him. They went to a local Starbucks and she sat him down.

'She showed me a form that's hidden, but it still exists today. It's just not indexed on Google or really shown publicly. There is a special form that you fill out and it went to a different set of administrators who use, in my opinion, more critical thinking skills.'

Boyd got the review taken down and, in return, he got a check from the doctor that at a stroke lifted him out of debt .

'When I got that check from the doctor, I'm like, "Holy moly!" – like a new world just kind of opened.'

For Curtis Boyd, this was the introduction to a world of ratings wars. He ended up creating the Transparency Company and we'll meet him again in Chapter Eight, camped on the frontline.

Boyd's experience shows two things: one, that reviews *can* be taken down, and two, that there's money to be made in taking them down. Ironically, sites like RateMDs have twigged as much: they are happy to make you pay to clean up the reputation you already paid to have sullied.

In 2018, Dr Sukhbir Singh, a gynaecologist at the Ottawa Hospital in Ontario, Canada, found an anonymous posting on his RateMDs page from a patient claiming that he had harmed them during a medical procedure. There was just one problem: it was a procedure that Dr Singh didn't perform.

Singh quickly responded on the site, advising the person to speak to the hospital about their concerns. But then, as quickly as it had appeared, the post vanished.

'This is crazy. None of this makes sense to me,' he told CBC News in Canada. 'I feel this is akin to cyberbullying.'[8]

What was crazier was that soon after Singh had responded to the review, a sales rep from RateMDs contacted him offering 'reputation management tools' for a fee. The 'service' included the ability to hide up to three comments from public view.

'That just made me sick to my stomach,' he said. 'It doesn't seem that in a public healthcare system that I should be marketing myself, that I should be protecting my reputation and paying an independent private company to do that work.'

To this, many doctors would say, 'Welcome to the starry future, Dr Singh.' There are now myriad paid services that offer to get reviews taken down. Removify, for example, offers 'fast, effective removal of RateMDs reviews' on the grounds that 'without any system for verifying genuine patient experiences, and confidentiality laws preventing you from effectively defending yourself with details, unfair and inaccurate reviews are a cancer on your reputation [nice work, Removify] and should be treated as such. The most effective cure for unfair reviews is fast, permanent removal.'[9]

But if you can 'removify' your reputation anyway, like surgery on a carbuncle, then the whole purpose of reviews from the consumer's point of view – clarity, feedback, honesty, trust – becomes moot. The journalist Kelly Crowe, writing for CBC News, discovered that RateMDs was offering doctors two paid-for plans to promote themselves on the site. The 'Promoted' package cost $179 a month and included the ability to run your own ads on other doctors' pages. But, wait, because for $359 per month, a doctor could buy the 'Promoted plus' option. Both packages allowed doctors to hide up to three unfavourable comments ... until the doctor stopped subscribing to the promoted packages, at which points the bad ratings reappeared.[10]

The solution to all of this, as it is with Amazon and Trustpilot, comes in verifying reviews. In the UK, Doctify is the major review platform for the country's healthcare industry, covering both private and NHS specialists.* It claims to be 'uniting the global healthcare community, one trusted connection at a time'.[11] Doctify calls itself 'Tripadvisor for Doctors', as if a hip replacement is the same as a weekend in Bruges, and it is a closed review platform on which all feedback is verified.

'We only ever want to publish genuine experiences from real patients,' says the site's co-founder Suman Saha. 'Before we launched Doctify, we noticed just how difficult it was to find reliable, representative and verified healthcare reviews. Neither patient nor provider could get easy access to feedback they felt confident in.'

Doctify's aim is noble, but its output is still star ratings and amassing reviews. The problem is, as usual, double-edged. Either you let everyone say what they want to say and receive a torrent of abuse and bile, or you 'close' the review site, let the doctors get involved and as such only get four- or five-star reviews. This makes the reviews redundant for end users because, self-evidently, not every doctor is unimpeachable.

* You'd have thought that, on the NHS, you get the specialist you're given and you're thankful for it, but NHS Right to Choose (RTC), introduced in 2008 in England, gives patients the legal right to pick their provider and clinical team for a first outpatient appointment, often via private providers contracted by the NHS.

Among the doctors I spoke to in the UK, Doctify is seen as a useful tool if used wisely. The supposition is that readers can spot a nutter patient with an axe to grind, if the site hasn't stymied them already. But all of the medical review sites demand that doctors pay to be listed on them. At this point, they cease to be independent arbiters and become marketing services. This is a contradiction that cannot be unpicked.

All of this suggests why the more senior doctors I spoke to try to stay off RateMDs and Doctify altogether. 'I do know some people who sort of say, "Please could you review me?"' said one London consultant who works in both the NHS and private practice.* 'But it would fill me with such hatred to ever have to stoop so low that I would never, ever, ever, ever ask. So I've always tried to avoid it.'

What this all comes down to is whether a service like Doctify does actually identify the best doctors. Since 2023, it has handed out Patient Voice awards and Doctor of the Year awards.

'We see this as an exciting step forward for Doctify and the sector as a whole,' said Suman Saha when the awards were announced. 'With the new word of mouth online, we believe it's increasingly important to give patients a voice and elevate how the sector captures and uses that voice in the provision of care. We see these "awards" as the first step in increasingly recognising great care in a transparent and validated manner.'

It depends who you ask. One former winner, who is currently rated as near as dammit to five stars after several hundred reviews, gets a very different reception from their peers.

'Everyone but everyone thinks they are a shit surgeon,' said a

---

* This is commonplace in UK medicine. Most doctors are employed by the NHS but, as they become more senior, they are allowed to set up in private practice as well.

contemporary who works in the same field. 'It is a brilliant example of where the person is so gormless that they would *ask* people to give them a Doctify review.'

To determine whether Doctify or the person I spoke to is correct, you'd have to let someone go at you with a scalpel. Ultimately, you'd still have to trust them, Doctify or not. It's why when star ratings get personal, we enter disputed territory. As all of these sad and angry vets, doctors, lawyers and cleaners show, it can get weird, nasty and, in extreme cases, be damaging.

★★

Weird, nasty and damaging is, of course, ideal territory for entertainment. A 2016 episode of *Black Mirror*, Charlie Brooker's dystopian TV looking-glass, took things to their logical conclusion, as that series tends to do. In 'Nosedive', we land in a near-future society in which everyone shares their daily activities through eye implants and mobile devices, rating their interactions with others on a scale from – you guessed it – one to five stars.

The consequences of such rampant rating are shown in all their socio-economic misery. The protagonist of 'Nosedive' is Lacie, played by Bryce Dallas Howard.

'Lacie is lost in this world where she thinks her worth is equivalent to her value in points,' said Dallas Howard when the series launched. As the story progresses, we see that if Lacie can just get from a 4.2 to a 4.5, she'll get a discount on a luxury apartment. Interacting with people with higher ratings raises Lacie's own score, but causing a scene at an airport when her flight is cancelled gets her 'double damage' and she loses an entire point for 24 hours.

It's a social sucker punch, one that leads to Lacie's rating spiralling and her winding up in prison with her eye implants removed.

In a wry coda, this low-point turns out to be a glorious release. Lacie and the man in the cell opposite realise they can now speak freely without worrying about being rated and, as the credits roll, they happily hurl insults at each other like a food fight.

'"Nosedive" is a satire on acceptance and the image of ourselves we like to portray to others,' said Brooker at the time. 'Everyone is a bit heightened and false because everyone is terrified of being marked down, because the consequences of that are unpleasant. So basically [it's] the world we live in.'

I spoke to Brooker about 'Nosedive' a decade on.

'The original idea was slightly different,' he says. 'It was going to be a *Brewster's Millions*-type story where the idea was that there was somebody who had the highest rating in the country who gets blackmailed into having to reduce their rating to zero. So this character runs around doing increasingly offensive things and then, in a slightly Charlie Sheen way, the more outrageously they behaved, the more popular they became – and so the problem was that their rating kept going up.'

Brooker couldn't make that story scan. 'I couldn't work out the mechanics really, because once you'd set it in motion, you just had a story where somebody was doing increasingly offensive things one after another.' But in playing around with the notion of societal rating, he hit upon the idea for 'Nosedive'.

'I'd read something about waiting staff being ranked out of five for something or other and that that was determining what shifts they got. And I thought, *Oh, that's fucking horrible. That you could be marked down because you were depressed that day or somebody just didn't like your face.* And so it span out of that.'

Booker wrote it at around the time that everyone started taking Ubers instead of taxis, when Uber's two-way rating system was making headlines.

'It was the notion that Uber was rating the passengers too, I think. That was sort of new information to me at the time and it got me thinking.'

When he saw the first cut, Brooker was worried that 'Nosedive', where the world is pastel-hued and striving for a Barbie-fied perfection, was too light. Up until that point, *Black Mirror* had made its name as a bleak, dystopian techno-futuristic fiction. So he showed it to a friend in advance.

'About five minutes in, she said, "This is an absolute fucking nightmare." And I was like, "Oh ... yeah." But then we agreed that it was a nightmare because "Nosedive" is what the world's like.'

As with any science fiction, 'Nosedive' takes modern concerns, dusts them with a crisp seasoning of paranoia and feeds them back to us. But that doesn't mean those concerns shouldn't be concerning. Brooker says that the episode was about 'the gamification of social interaction, basically'.

'I was arguing that social media is a video game in which you play a character and you press lettered buttons in order to accumulate points and followers. It's all about your reputation and your score. In a world in which everything is ranked and rated all the time, you're performing ... everyone's performing to some degree. But when there's ratings and a score, it's a sort of receipt. There's a constant sort of gauge and it warps everything somehow.'

I tell him about this book and about how it makes 'Nosedive' seem eerily prescient.

'I don't know if it was so much *prescient*,' he counters, 'because it was reflecting something that already existed. But it wasn't maybe as widespread as it has subsequently become.'

A British reality show from 2018 called *The Circle* took what Brooker calls 'the gamification of social interaction' to another level, by literally making social interaction into a game.[12] Channel 4's social media-age spin on *Big Brother* put a group of contestants in a block of flats, but never let them meet face to face. Instead, they communicated through messaging devices, allowing them to adopt whichever persona they liked. Crucially, the players rated each other throughout the game, from one to five stars with average scores revealed at the end of each 'rating'. Eliminations (or 'blockings') occurred when selected players, usually the highest rated in a rating, became 'influencers'. The influencers determined who was up for elimination in the next episode. Just as with the (fictional) 'Nosedive', in the (heavily factional) *The Circle*, your star rating determined your future prospects.

Tim Harcourt, the creative director of Studio Lambert and a reality TV giant who has also produced global hits like *Gogglebox, Race Across the World* and *The Traitors*, came up with the idea.

'The starting point for *The Circle* was to do a reality show where people never meet. Then the second idea was is there a reality show that's a bit like *Rear Window* [the 1954 Hitchcock classic shot almost entirely from within a single room and taking a single point of view looking out the window]? I know that that sounds quite highfalutin, but I was playing around with these ideas like how can we do something where everyone is living in an apartment but you don't quite know who they are ... and they never meet.'

Channel 4 were looking for a show about social media. Harcourt combined his idea with theirs and came up with a social media competition where people never encounter each other in person.

As part of his research for the show, Harcourt and his team looked for precedents. He looked at dating apps, at Brooker's 'Nosedive' episode, at RateMyPoo.com*, at the Chinese social credit system (see later) and at the website Hot or Not.†

---

* Rate My Poo was a shock site that went live in 2001. The site allowed users to upload pictures of their turds so that other people could rate their

In particular, they watched a 2005 series called *Hot Tub Ranking*. This was a brutal and much-derided UK reality show in which *Big Brother* creator Endemol gathered a group of semi-naked women and got them to bitch about who has the best body, while surreptitiously being ranked from one to five in order of 'sexiness' by a group of men secreted behind a two-way mirror. It transferred to the US in the form of *Battle of the Bods*.[13]

'I think I was in denial that we were making a popularity contest at first,' Harcourt says. 'I kept on saying, "Oh, it's a game of communication", but eventually I sort of had to go, "Well, it is effectively a game of popularity. It's like who's most popular when they have to operate behind the mask or behind the screen."

'And then, when I realised that, it sort of unlocked quite a lot. It then made the whole ranking and the rating much more powerful, and that actually made me realise what we had to do with it was to create winners and then losers ... and the winners had the power of life and death over the losers.'

What Harcourt had hit upon was that when it comes to rating individuals, not products or businesses, a rating *is* a ranking. It's not just a classification; it's a competition.

'Most of the contestants in *The Circle* were under 27, 28, so for them it was just second nature to give things five stars, four stars, one star.'

---

creations on a scale from one to ten. In 2019, the site shut down and all pages redirected to an image of the (then former) president Donald Trump. His picture was captioned 'Biggest Piece of Shit'. There were multiple variations of Rate My Poo, such as Rate My Kitten, Rate My Finger and Rate My Puppy. Ratemypoo.com now redirects to ratemypussy.com. Which is nice.

† As the names suggests, a popular early internet rating site that allowed users to submit photos for others to rate. Discussed below.

Channel 4 were worried that the contestants were going to find it hard to rank everyone around them. They didn't at all. For their generation participating, rating and ranking had become part of everyday life. But they soon hit a snag.

'In the first series, we used star ratings, but in later series we changed to rankings. What we quickly realised with the star ratings was that one person might say, "I really fancy that girl, I'll give her four out of five", whereas another person might say, "I really fancy that person, I'll give them five out of five."'

For series one of *The Circle*, this led to what Harcourt calls 'an impure result'.

'The guy who probably should have won came second because a friend of his gave him four stars, not five. I remember thinking at the time that that's a real problem with ratings. They mean different things to different people. My dad might have a great Uber ride and go, "Right, that's four stars, really good", but a mere four stars is death to an Uber driver.'

For the second series, they changed the ratings system to a ranking system. 'It was a bit more uniform or less relative across the nine or ten players,' says Harcourt, 'and much better in terms of the narrative. It's so much better to be able to say who's your favourite person and who's your second favourite person.'

It worked better as entertainment, too. In the US, Netflix picked up *The Circle* for an American version, which ran for three series.

★★★

Both *The Circle* and the 'Nosedive' episode of *Black Mirror* were TV concoctions (one UK reviewer gave *The Circle* two stars himself and labelled it 'a vast maelstrom of species shame'[14]), but, as with all such concoctions, they had footings in reality.

Peeple, for example, was a mobile application launched in 2015 with a name that was probably more telling than it was meant to be, suggesting both mass-market appeal with just a soupçon of stupidity and creepy surveillance. The app encouraged people/peeple to leave recommendations for other people based on whatever working, romantic or everyday interactions they'd had with them. Initially described as a 'Yelp for People,' its launch announcement in October 2015 made a splash – but perhaps not as its makers intended.

'Wait, so Peeple is a forum to abuse and slander anyone on the internet? We already have that. It's called the internet' was one wry response on what was then called Twitter.[15]

The app attracted criticism over concerns of harassment, leading its creators to launch a bowdlerised version of Peeple in March 2016. Notably, Peeple v1.1 omitted star ratings, even though they had been present in the (generally lambasted) beta.

Lulu was a mobile app from 2013 that described itself as 'a private network for girls to express and share their opinions openly and honestly'. Specifically, those opinions were about men – what they looked like, how much money they had and whether they could be trusted. In essence it was an app-based *Hot Tub Ranking* for boys, and it gave a ten-point score to whichever sorry Y-chromosomes happened to fall under its gaze.

Neither of these apps lasted long. In both cases, the court of public opinion ruled unequivocally that rating people like they did in 'Nosedive' (in the way that we all rate each other but without sharing it publicly) was beyond the pale. In part, this was because both apps quickly regressed to what appears to be the rating mean: slagging people off, quick-fire defamation or digital revenge.

Peeple even prompted one UK columnist to opine that 'the cult of ratings is destroying us, one star at a time. If I had to rate Peeple – and I really would like not to – it would go right down there, below

Amazon's taste profiling, iTunes' stars, RateMyFace and RateMyPoo. The only possible reason why anyone would sign up to Peeple has to be to engage in a new exploration of misanthropy and to find a new way to create a cocoon of miasmic boredom in which he (or she) can hide away and hate in abject misery.'[16]

It is telling that, when couched in the language of commerce, rating people is not just acceptable but desirable and ubiquitous, as we have noted. Does the seller deliver on time? Did the product work? Was the item as described? Yet when ratings cut to the quick – is this person a cad, a chancer or a crook? Do we, in short, like them? Should we trust them? – ratings and ratings apps are suddenly reframed as creepy tools for harassment and manipulation. In short, they get a bit 'Nosedive'-y.

One reason this kind of first impressionism is unsettling – powerful in TV drama and entertainment, but disconcerting in real apps used by real people – is because it is actually deeply human. The brain groups faces by gender or social status or race with astonishing speed – around 50 milliseconds of exposure activates the amygdala.*[17] The evolutionary aim is to divide the world into Us and Them groupings, on the understanding that the odds of survival increase if you hang (and mate) with the former. The hormone oxytocin is released during these interactions, which prompts trust, cooperation and generosity to whomever you deem an Us (and the opposite to the poor Thems). It's unconscious, ages-old and automatic.

* That is half of the literal blink of an eye, which takes on average a tenth of a second or 100 milliseconds. This creates an emotional reaction before the fusiform face area (the cortical region that deals with facial recognition) categorises whoever you're looking at and tells you not to think socially unacceptable things like, 'yuck, they're gross'. It's an unconscious bias.

It's revealing that apps that have the temerity to cut to this neuro-evolutionary chase – Who's hot? Who's nice? Who's a total idiot? – have been stigmatised, given that so much of human interaction is about precisely this kind of brutal, snap assessment.

Yet, at a certain point, skin-crawlingly honest appraisal ('I don't like you because you have big ears and you smell and I just don't like you'), which is hard-wired into our ape-brains, is deemed morally reprehensible. But because it is hard-wired into our brains, we can't help ourselves from doing it. There's *The Circle*.

Dating apps are a crucible for this particular uncomfortable truth. Rating people directly for their attractiveness, with all the many connotations of that word, is the crux of dating apps. They just don't like to talk about it. Be it Badoo, Bumble, Hinge, Match, POF or OKCupid, it is fascinating to see how, in their PR, all of the many successful apps chose to skip across one crucial fact like bare toes on hot coals: dating apps *are* rating apps.

On Tinder, for example, the biggest and most-used dating app worldwide, they say they are able to 'make the impossible possible' – that is, find you a perfect match – thanks to their 'fancy numbers', by which they mean their algorithm.

But when it comes to the method behind their matching, Tinder go a little sheepish. They say they use 'anonymised cues from photos to help tailor recommendations ... We are constantly honing the potential matches users see based on how often their profile – and all profiles in their area – are Liked or Noped.'[18] Yet liking and noping – swiping one way or the other – is a binary rating system.

Tinder used to use an Elo score system.* This is the same method used to calculate the skill levels of chess players: your ranking improved

* Named after its creator Arpad Elo, a Hungarian–American chess master and physics professor.

if lots of people swiped right ('liked' you), but how much it improved was based on who was doing the swiping, just as beating a chess grandmaster counts for more than beating a five-year-old. Being liked by someone who was popular already would add more to your rating.[19]

But now that Tinder has its own near-bottomless data pool, it has done away with the Elo system, announcing in 2019 that 'Elo is old news at Tinder. It's an outdated measure and our cutting-edge technology no longer relies on it. Today, we have a dynamic system that continuously factors in how you're engaging with others on Tinder through Likes, Nopes, and what's on users' profiles.'[20]

If you actually have a go on Tinder, or just watch someone else in the act of seeking a soulmate, you soon find that whatever its secret-sauce 'dynamic system' is, its user interface certainly appears to remain heavily reliant on people's physical appearance. And if you check across Bumble, Badoo, Hinge, Match, POF, OKCupid, eHarmony and the rest, it is plain to see that looks have never been, and will never be, an outdated measure.

All of them are rating individuals – they just don't like to say so. This moral ick goes back to the early days of the internet. In the late '90s and early '00s, there was a profusion of sites like HotOrNot, Buddypic, FaceTheJury, Blingshot, Realpics, Cpixel, The Dilly, Picturepoll, doyoulookgood.com, findapix, makeoutclub, becauseimhot, ratemyface and so on. They all relied on the same concept: asking users to upload their picture and then let other users give them a numerical rating.[21]

The history of Hot Or Not, the most prominent of these self-worth immolation devices, is instructive. Launched in October 2000 by two electrical engineers at the University of California, Berkeley, it was an instant hit. Two million people were on the site every day inside its first week and, within a few months it was one of the top 25 domains on the entire internet.

It was cultural winds, not commercial ones, that sent Hot Or Not off course, as people began to suggest that maybe judging individuals solely on their looks was a little degrading. Tellingly, however, Hot Or Not evolved into a matchmaking site and was eventually bought by Badoo, now a huge dating site.

When co-founders Chad Hurley and Steve Chen first started work on YouTube, they were trying to make a version of Hot or Not with video.[22] Rating and ranking based on what used to be called the passions – lust, anger, etc – is baked into the internet, just as it is baked into our brains.

★★★★

Tech guys love algorithmicising things like lust and love and awe and music. Assigning numbers and process to things that arts graduates thought were matters for poetry and the soul (and thus ineffable) is in many ways the backbone of the digital project.

It's also a mark of totalitarianism overreach, because the ability to measure and assess offers the panacea of mass control. In 2014, China introduced its infamous Social Credit System, which appeared to be state-sanctioned 'Nosedive'. The system, at least as it was reported in the West, was Orwellian and deeply chilling.

The billionaire philanthropist George Soros used his annual dinner at the World Economic Forum in Davos five years later to call out China. 'I want to call attention to the mortal danger facing open societies from the instruments of control that machine learning and artificial intelligence can put in the hands of repressive regimes. I'll focus on China, where Xi Jinping wants a one-party state to reign supreme.'

The former hedge fund dealer warmed to his theme. In China, he went on, 'all the rapidly expanding information available about a person is going to be consolidated in a centralised database to create a "social credit system". Based on that data, people will be evaluated by algorithms that will determine whether they pose a threat to the one-party state. People will then be treated accordingly.'[23]

In fact, China's Social Credit System came from a similar starting point as our own star rating and review system: a need to somehow quantify trust. Its origins were in the exact same financial credit system that the West had developed around a century earlier. Financial credit and money-lending has been around as long as there has been trade, but local credit bureaus first emerged in the West in the late 19th century to share borrowers' information. The problem was that their judgements were often subjective – they'd pay more attention to a loanee's moral character and whether they liked them instead of the ledger of what money they'd paid back on time in the past.[24]

As consumer credit exploded throughout the early 20th century, a more robust, less biased system was required. In 1956, two American engineers, Bill Fair and Earl Isaac, created the first statistical models to predict credit risk. Fair, Isaac and Company gave their initials to what is now known across the US as the FICO score. (In the UK, Experian, Equifax and TransUnion do much the same thing but we don't have a snappy acronym for it.)

In many ways, China has followed a similar path, but a generation or two later. In the 1980s and 1990s, rural Chinese citizens looking to take out mortgages or start small businesses rarely possessed formal credit histories or documentation. This made it difficult for banks to know whether or not to lend to them, and the lack of a paper trail made financial regulation difficult as well.

In 1999, the Chinese Academy of Social Sciences produced a report entitled 'National Credit Management System'. The then Chinese premier Zhu Rongji ordered the People's Bank of China (PBOC) to begin work studying credit systems. Again, this was in response to growing concerns about our old friend 'trust' in business transactions, allied to a growth in defaults, tax evasion and other financial flim-flam.[25] Trust, credit (in its original sense) and control are at the root of all ratings systems.

Financial credit ratings soon morphed into social ratings, adding a broader, creepier moral dimension. Some provinces introduced a local points scheme, while others began to assess public behaviour and law abidance as well. The semantics of credit were being explored as much as people's financial propriety.

The burgeoning private sector also got involved. From about 2015, companies like Alibaba, with their 'Zhima Credit' (also known as Sesame Credit), built credit scoring products for individual users by looking at both their financial behaviour (payments, loans) and their non-financial data (shopping, what they did online). These private systems were all different and they weren't formally part of the government's system, but they nonetheless planted the seed of an everything rating for everyone, all of the time.

The big year was 2014. It was when the Chinese State Council issued the catchily titled Planning Outline for the Construction of a Social Credit System (2014–2020).[26] (I would have gone with Rate-U-Like, although I'm not sure how that would translate to Mandarin.)

This document is well worth a read if you're a fan of dystopian sci-fi. It's probably best read in an ice-bath. It is written in the sort of stern-faced technocratese that Orwell could scarcely better (though I imagine it loses a bit of life in translation).

'Perfecting the social credit system is a necessary condition to deepen international cooperation and exchange, establishing international brands and reputations, reducing foreign-related transaction costs, and improving the country's soft power and international influence, and is an urgent requirement to promote the establishment of an objective, fair, reasonable and balanced international credit rating system, to adapt to the new circumstances of globalization, and master new globalized structures.'

The aim of the document, if you can get through it, which is unlikely unless you've been heavily sedated, is to formalise the government's intention to build a unified, nationwide system. Credit information collection, supervision, regulation and reward will be standardised! A credit information infrastructure will cover all of society, there will be a market for credit-services; and crucially there must be a reward/punishment system for keeping or breaking trust! And there was a sort of deadline – by 2020, the system was to be well-established nationwide.

Resistance is futile, humanoids. Over to you, Charlie Brooker.

This was the document that got George Soros's back up and, reading it, you can see why. But the important thing to register about the Chinese Social Credit System as it has actually played out is that it is *not* 'Nosedive'. It is not a single, Orwellian, dehumanising point score that determines your options and outlook. There are lots of different systems operating across different jurisdictions and spanning the public and the private realm. As such, the 'system' is not really a system, in that it is neither unified nor consistently enforced.[27] Different provinces, cities and companies have different pilot schemes; some regions are stricter, others less so. Some systems remain local and fragmented.

And so the most obvious criticism is that to this day it is hard to say exactly what the Chinese Social Credit System is at all. What data is used, how decisions are made and how appeals work are often opaque. Questions about accuracy and fairness and how flustered we should all be by the tentacles of the Chinese state asphyxiating all human individuality are therefore often predicated on not really knowing what's going on.[28]

Read that last paragraph back and the irony squirms off the page: China's control system, a Communist tool supposedly for social governance, is not worlds away from where we have landed up in the late-capitalist West. As shown above, we invented financial credit scores. Big tech knows just as much about you from your online behaviour and purchases, your keystrokes and locations and Alexa commands, as the Chinese state knows about its citizens. Do the top-secret algorithms of the companies that run our world end up assigning you a single score? Probably not, but they are undoubtedly using numbers to grade you in all manner of ways. Even filing a star review is unwittingly handing over hundreds of data points to a company that may well have 'trust' in its name.

In the era of what the writer Cory Doctorow calls 'enshittification', we now know that Big Tech's early evangelism and shiny motives have been overtaken by the appetites of hyper-capitalism. If you can quantify the unquantifiable, you can sell it. If you can rank people, you can control them – where control can mean controlling their wallets and their choices as well as their thoughts, proclivities and beliefs.

In other words, if you find China's attempts at rating people creepy, well, look around you. From your financial credit rating to your every interaction with the internet, you are constantly being rated yourself.

It's worth remembering that – just like YouTube – Facebook, the dubious grandaddy of all modern social media, began life as a hot chicks rating app for Mark Zuckerberg and his friends at Harvard. In later years, it would lay down much of the foundation for the digital economy by noting who was doing the rating as much as how they rated things, and then selling that information to advertisers. In effect, its algorithm rated you.

What Zuckerberg had hit upon was that rating people hot or not was an unpalatable but deep and unerasable human trait, and that how people rated other people was itself a way of rating them. It was all a giant wheel, a big circle of reflective and representative intel.

Facebook's great leap forward was to realise that all of this information could be monetised. And lo, Zuckerberg went forth and monetised uncontrollably. Meta – a business name that literally means self-referential – is now the sixth-largest company in the world and Zuckerberg, by playing God, has become richer than God.

All of it began, though, with an app for rating individuals.

And what's wrong with that? We have been rating individuals for years. It's called voting. This is where people ask to be rated, expect to be slighted and accept that rejection will hurt, but where the prize is that other intrinsic human desire: power.

And that is why we're going to Oregon.

★★★★★

Chapter Six

Have Your Say

'The private citizen today has come to feel rather like a deaf spectator in the back row ... He does not know for certain what is going on, or who is doing it, or where he is being carried. No newspaper reports his environment so that he can grasp it; no school has taught him how to imagine it; his ideals, often, do not fit with it; listening to speeches, uttering opinions and voting do not, he finds, enable him to govern it. He lives in a world which he cannot see, does not understand and is unable to direct.'

Walter Lippmann, *The Phantom Public*, 1925[1]

That was the American journalist Walter Lippmann writing more than a century ago and it is hard to argue that, for the private citizen today, things have got anything other than considerably worse. In the nuclear winter of the internet, we are not merely deaf spectators to public events – we are punch-drunk on hot takes from social media, high on the hallucinogenic hoodoo of politicians who should know better and roundly stupefied by the amount of facts and alternative facts (and the very idea that there may be alternative facts) being fired at us from our beloved devices. It all makes the question 'What do you think?' more complicated than it has ever been.

In modern democracies, the principal time when you're asked to say what you think, and when you're assured that what you think matters, is when it comes to voting. Unfortunately, these are not good times for the primacy of the vote. In October 2025, both the German chancellor Friedrich Merz and the French president Emmanuel Macron made speeches at the same event talking of the 'degeneration of democracy'. On the one hand, they blamed external factors – demagoguery, exploitative politicians, manipulative elites – but they also both acknowledged that democracy itself was floundering.

'On the inside, we are turning on ourselves. We doubt our own democracy,' Macron said. 'We see everywhere that something is happening to our democratic fabric. Democratic debate is turning into a debate of hatred. Belief that democracy is the form of governance best suited to the modern world is dwindling.'[2]

In the US, a 2025 survey found a record 64 per cent of Americans believed their democracy was 'too politically divided to solve the nation's problems'.[3] In the UK, a comparable poll of people aged 16–29 found 63 per cent thought democracy was in trouble. While young people 'preferred to live in a democracy over a dictatorship' by 57 per cent to 27 per cent, only 35 per cent said they would consider getting involved in organised politics.[4]

This amounts to a vote of no confidence in voting itself. There is a loss of faith in democratic systems. We think our vote doesn't count. We think it doesn't matter.

★

If people in Western democracies are starting to believe their vote doesn't matter, this is often because their vote *doesn't* matter. The maths behind most of the world's voting systems means import tends to reside only in certain states or seats. If you're not in one of those states or seats, you're immediately one of Walter Lippmann's deaf spectators, fumbling around in the back row.

We read all the time about voters who feel left out or ignored. In the UK, the 2016 Brexit vote was won with a slogan of 'Take Back Control', implying that people felt strongly that they had somehow lost control – and that therefore someone else was in control. It is a state of mind that, no matter how illusory, engenders a perpetual state of niggling impotence. Something has to change, it niggles. Might as well give [insert bumptious extremist here] a chance – how much worse could they be?

The British political writer James Ball calls this the rise of the 'fuck you' party, a break-glass-in-emergency button pressed by the exhausted and the indifferent, just to see what happens. It is the presiding mood of Western politics in the first quarter of the 21st century and history tells us that it won't end well; those who feel disenfranchised and tongue-tied tend to sway towards demagogues, populists and the far right.

At the base of this sense of impotence is a feeling of not being listened to. A 2025 study by the Policy Institute at King's College London, Ipsos UK and the London School of Economics found that while large majorities of people in the UK have long felt powerless to influence decisions both nationally and locally, there has been a 'big decline in the belief that public services are listening'.

Professor Bobby Duffy, director of the Policy Institute, qualifies this, saying that 'there was never a "golden age" of the public feeling particularly powerful in their influence over national and local decisions'. But he's still no Pollyanna. 'We are seeing declines even from that low base. Now only a quarter believe that getting involved in politics can change things, down from a third in 2018.'

The same report showed that we've also seen a collapse in the belief that public services are ready to listen to complaints, halving since 2002. We know that in commerce how companies respond when things go wrong is almost more important than how they make things go right. In government, they call this 'delivery' and in both the UK and the US, it's a crucial component in the battle for the idea that government works at all. According to Professor Duffy, results like those in the survey above, where people feel that public services just aren't listening to their issues, 'show how difficult a task [governments] face in convincing the public that services can actually deliver for them'.[5]

★★

And yet. Every morning in your inbox, you more than likely get several emails telling you 'We're listening' or 'Your opinion matters,' or that right here, right now, you can, 'have your say'. Beside it is a star rating followed by a review form that aims to prove as much. Fill it in and you get another email or SMS thanking you, possibly even a written response saying that your feedback has 'made a difference'.

Someone, even if it's a robot, *is* listening and, for a second or two, you are gifted a sense of agency and power. You might even want to press the 'fuck you' button and lay into some major corporation that you feel has let you down. Either way, go on, rate something you've just bought or a service you've just used – at least *that* vote matters.

Note the language – 'trust', 'faith', 'engagement', 'complaints', 'responses', 'listening' ... The lexicon of customer feedback is the same as that of modern functional government. Here then is a forum in which everyone is voting every day, even at a time when we're told no one is interested in voting. The surprise is that as people wring their hands asking how in heaven's name we get young people to commit to an opinion on the issues that matter, that generations are being schooled in what is, at base, a shadow play of democracy. We are all voting, rating and ranking pretty much continuously. It's just that what we're voting on is what we thought of our matcha latte this morning or whether those new running shorts really 'delivered' on quality and price.

Why not cross the streams? Why not use people's inherent desire to have their say and be heard to encourage them to have their say on something that really matters? Around the world, almost everyone agrees that their voting system is rubbish, but no one agrees on how to make it better. It's amazing that no one has thought of trying star ratings in politics.

Actually, they have. In the state of Oregon, they've put two and two together and made five (stars). The impeccably-titled-for-my-purposes STAR voting system is an acronym for 'Score Then Automatic Runoff'. You rate your favourite candidate on the well-trodden scale of zero to five, or you can leave them blank and rate someone else. Equal scores are fine too – the aim is to register a level of support.

Next comes the 'Automatic Runoff' part of the STAR acronym. The two highest-scoring candidates go into a final. The humble voter needs do nothing, because their ballot already shows their preferences and so their full vote automatically goes to the finalist they prefer. The finalist with the most votes wins. As the STAR voting slogan goes, 'Add up the stars, then add up the votes!'[6]

I sensed you glazing over there in the battle for supremacy between your two-star and your three-star local councillor and I agree – no amount of cajolery or exclamation marks can camouflage the fact that this sounds much more complicated than 'vote for the one you like most'. I promise you it isn't. To prove it, let's have a quick go at what the people who invented it call a 'five-star ballot'. We'll go straight in on one of the most pressing issues of the day: which is the best pet? (The answer, obviously, is my dog, but don't let that sway your vote, chinchilla fans.)

If you go here: https://bettervoting.com/meta_pets or use this QR code:

You can play out a star-rating voting scenario. The poll to elect president pet will consist of the classic battle royale between Bird, Cat, Python, Dog, Fish, Rabbit and the easily-underestimated Rat. Off you go.

Welcome back. I don't know about you (although I will once you've all voted), but what I found was that because we collectively fill in hundreds of similar forms like this each day, the great pet referendum felt very easy. You'll have noticed already that this five-star ballot is in fact a six-bracket system, because you can no-star a candidate/pet with a 0. But you'll also have noticed several reasons why STAR Voting appeals ahead of a standard, choose-one-candidate ballot paper. Leaving aside the 'electability' fallacy (not voting for people you've been told can't win, for instance, Senator Rabbit, with the implication that you're wasting your vote)*, the main reason is that people are more likely to do it. And that's because we do stuff like this all the time.

---

* In the run-up to the 2020 presidential primary in the US, an Avalanche Insights poll cited by Nate Silver of 538, asked voters two questions: 'Who would you vote for if the 2020 Democratic primary was today?' and 'Who would you pick if you had a magic wand?' The different answers were telling. There was a 12 per cent uptick for the *perceived* front-runner

This means we're also more likely to do it right. I have a friend (it's actually my wife) who works in the public sector and regularly helps with local and general election counts. She tells me that the number of scrawled penises on the ballot papers never fails to delight and amaze. Obviously, there's no situation in life where a scrawled dong in a formal scenario is not hilarious, but away from the comedy, spoiled ballots are a real problem in elections worldwide. The single-mark ballots used for most elections in the US, for example, have a spoilage rate of between 1 and 4 per cent. Rated ballots outperform single-mark ballots on this metric with a spoilage rate of roughly between 0 and 2 per cent. At this point, no one is quite sure why. One possibility is that voters pay more attention when they have to express a preference for each and every candidate, reducing the rate at which they make mistakes. My contention is that they have got very good at rating things on a scale of one to five.[7]

★★★

STAR Voting was developed in 2014 by a US non-profit called the Equal Vote Coalition. As their name suggests, the EVC are unhappy that, under current voting systems, all votes are not equal: strategic voting means that a lot of people always end up voting for someone they might not actually like in order to stop someone they definitely hate. It feeds into a system fuelled by negatives rather than positives, as well as one in which people feel their votes don't matter ... because, yet again, in many cases they don't.

'In this time of unprecedented polarization,' the EVC told me, 'STAR Voting's 0–5 star ballot transcends the yes/no,

---

compared to the next closest – despite the fact that he would have been tied for second place if people had just voted for who they actually liked.

us-against-them narrative. The result is a system that empowers voters to vote their conscience, levels the playing field and elects more representative candidates.'

What about my idea that star rating might work for voting because we vote with five stars every day? Arend Peter Castelein is a production lead at the EVC.

'As we developed Star Voting, we really found that having the six columns, zero through five, was a sweet spot for people. It mapped well to cognitive load – the one thing you don't want to do is overwhelm voters. STAR Voting just all fits so elegantly together with the acronym and the name. It's intuitive to voters from the start.'

Not only that, but it also simplifies the electoral process and can therefore save candidates and taxpayers money.* (In the US, where primaries are used to pick candidates before elections, STAR Voting works like a primary and a general election all in one – in many cases voters could skip the primary and only one election would be needed to find a majority-supported winner. This means less lobbying, less campaigning, fewer voters getting bored of endless elections and more time for elected candidates to do the actual job of running the government.)

It sounds to me like just the ticket for the ticket. STAR Voting has been used in elections big and small already, from when the Independent Party of Oregon used it for their state-wide primary in 2020, to internal elections (such as when the Democratic Party of Oregon ran its presidential delegate elections on the system in the same year), to organisational elections, such as when the LA chapter

* It doesn't simplify the ballot *counting* process though, meaning we would lose the classic images of vote-counters next to piles of ballots racing to be first to declare on election night. I'm not sure this is that great-a loss.

of Democratic Socialists of America used STAR Voting for their ballot in 2024. It seems to work.

But back to the pets ... You'll note that, with this kind of voting system you assess each pet on its merits. You're not *ranking* them against one another, you're *rating* them individually – several pets can have the same rating. It's then up to the algorithm to determine which one gets the top collective score. This is big, because it's just like on Amazon. There, each review asks you only what you thought of each item discretely, not in terms of how the item compares to all the other similar items. Statisticians will tell you there's a big difference.

It makes sense to bring the language and ethos of consumer choice into voting, if only because we are all consumers and we have become very good at choosing. We know by now, when it comes to purchasing products and services, that we need to do our research, read other reviews and be suspicious of glowing endorsements as much as weaponised denigration before we click to buy. But we rarely do the same before parting with our vote at the ballot box.

In fact, virtually no one 'does their research' in this way when it comes to decisions that impact their nation's livelihood and future. In their classic study *Deliberation Day*, the American political scientists Bruce Ackerman and James S. Fishkin made a compelling case for how we might vote better, but first they made a compelling case for *why* we really need to.

'If six decades of modern public opinion research establish anything,' they wrote, 'it is that the general public's political ignorance is appalling by any standard.'[8]

Basically, very few of us have a clue what we're talking about when it comes to the weightiest matters in public life. In 1964, in the chilly depths of the Cold War, a majority of the American public did not know whether or not the Soviet Union was in NATO,[9] while in 2003 half of the country thought the 9/11 hijackers were Iraqis* when they weren't. This was the prevailing, erroneous opinion as the US government considered (and eventually embarked upon) a second Gulf War in Iraq.

In the UK, repeated surveys show that half the country doesn't know if we have a written or unwritten constitution. Very few people know the positions of the candidates they're voting for,† yet the collective outcomes of these votes affect all of us profoundly.[10]

Ackerman and Fishkin's proposal was a national holiday where we would all use the time to get together and talk about stuff. Deliberation Day, they called it (making it sound like the worst movie you've ever heard of).

'It will be held two weeks before major national elections,' they wrote. 'Registered voters will be called together in neighborhood meeting places, in small groups of fifteen, and larger groups of five hundred, to discuss the central issues raised by the campaign. Each deliberator will be paid $150 for the day's work of citizenship. To allow the rest of the workday world to proceed, the holiday will be a two-day affair and every citizen will have the right to take one day off to deliberate on the choices facing the nation.'[11]

It's a terrifying prospect, I know, but it was founded on their belief that voters weren't stupid; they were just too busy or too bamboozled

---

* None of them were. The group of 19 attackers came from Saudi Arabia, the United Arab Emirates, Lebanon and Egypt. They were chosen by al Qaeda leaders in Afghanistan in 2000 before undergoing training at various camps across the country.

† In the 2010 UK general election, a YouGov survey concluded that there was 'no evidence electors have their vote swayed by reading manifestos. Indeed, very few people bother to read them at all.' In 2017, a BMG Research poll said that two-thirds of the UK public either don't read manifestos, or they don't know what they are.

to get themselves well-informed (or even informed). It wasn't ignorance so much as apathy, or perhaps a combination of the two that I am calling 'ignapathy'.

If people could just do their research, they said, everything would change – the candidates, the media, the activists, the interest groups, the spin doctors, the advertisers, the pollsters, the fundraisers, the lobbyists and the political parties.

'All would have no choice but to adapt to a more attentive and informed public. When the election arrived, the people would speak with a better chance of knowing what they wanted and which candidates were more likely to pursue the popular mandate.'

Deliberation Day came out in 2004, some years before the rated society I've been chronicling became the norm. Now, people are deliberating in their droves all the time, before casting a 'vote' in the form of a star rating. The public can be attentive and informed – and the forces of ignapathy can be defeated – when they want to be.

What if ratings could be used before we even had to vote, to determine public policy as it was being made? The UK has been moving towards this kind of consumer feedback-led policy in recent years, albeit with mixed results.

In 2025, as part of a ten-year programme to reform the NHS, the prime minister, Sir Keir Starmer, announced plans for patients to determine how much hospitals are paid from central government by rating their treatment. Under the plans, patients would be contacted a few weeks after they'd been unfortunate enough to land up in hospital and asked if they thought their care was good enough for the hospital to receive full payment. If patients said no, about 10 per cent of 'standard payment rates' would be diverted to a local 'improvement fund'.[12]

British doctors weren't exactly happy at the idea of their funding being dependent on what was basically a star rating.

'None of our members have raised this idea with us as a way of improving care and, to our knowledge, no other healthcare system internationally adopts this model currently,' Matthew Taylor, chief executive of the NHS Confederation, said, adding that 'patient experience is determined by far more than their individual interaction with the clinician and so, unless this is very carefully designed and evaluated, there is a risk that providers could be penalised for more systemic issues'.

In other words, Taylor envisaged us being transported back to the world of RateMDs in the last chapter, where patients hammer doctors for all manner of reasons, justified or not. Whether they're just peeved or they have a genuine grievance, the upshot is that a hospital loses funding. That's bad for all users of the hospital.

It's not the first time that the UK has tried to use star ratings as part of public policy. In October 2001, the then secretary of state, Alan Milburn, announced the introduction of star ratings for social services.[13] All of the country's social services, covering vital areas such as care and support of troubled families, the protection of children at risk of harm and supporting people with disabilities, were to be assessed on a star rating from zero to three stars. The ratings had consequences, too.

'The best performing councils will have an increasing level of freedom in the way they use centrally provided grant funds,' said the Social Services Inspectorate. 'At the other end of the scale, councils with zero stars can expect increased levels of intervention, and will be subject to more rigorous and frequent monitoring.'

As it happens, my own local council, Haringey, was zero-starred. Very quickly, though, the whole idea of rating social services with stars at all was itself given a dire rating.

One academic paper, weighing up how it had gone, picked apart the methodology and cast doubt on 'the reliability and the accuracy of the measures used and the lack of consistency between "hard" measures and "softer" information (such as that from reviews and inspections) that together form the national judgment as to performance, employed at that time, by the Social Services Inspectorate'.[14]

This was an academia-speak way of saying people were comparing apples and oranges – that one person's three stars is not the same as another's. This is a starry issue we have been wrestling with since Chapter Two.

Star ratings in British social care lasted until 2009, but governments still can't get away from the idea of using all of this ranking and rating to assist in public policy. In the US, the Centers for Medicare & Medicaid Services (CMS) uses a 'Five-Star Quality Rating System' 'to help consumers, their families, and caregivers compare nursing homes more easily and to help identify areas about which you may want to ask questions'.

The official US government Care-Compare website allows potential users and their families to weigh up nursing homes via star ratings.[15] 'Nursing homes with 5 stars are considered to have much above average quality and nursing homes with 1 star are considered to have quality much below average. There is one Overall 5-star rating for each nursing home, and separate ratings for health inspections, staffing and quality measures.'[16]

The CMS star system quickly became ubiquitous, a popular way both for consumers to educate themselves and for nursing homes to attract new customers. During the coronavirus pandemic, with many locked-down homes unavailable for prospective residents or their families to see first-hand, the ratings were all anyone had to go on.

As always, however, the system was ripe for manipulation, mainly because it relied on self-reported data – and, as we all know, marking your own homework is a bust.

'They were working to improve their ratings, but not their quality,' said Charlene Harrington, who sat on a board that advised CMS on the ratings system. 'The problems with the five-star system left these homes less prepared in the pandemic,' she said, speaking to the *New York Times* in 2021. 'They were allowed to not have enough staffing, and they were allowed to ignore infection-control deficiencies, so they had poorer quality than the public knew about, and they were in the worst position to manage Covid.'

A system that promised to distil reams of information and transform an emotional process into one based on objective, government-sanctioned metrics had comprehensively backfired.[17]

In the UK, Ofsted, the national education watchdog, used to rate schools on a four-point ratings system – until a head teacher, Ruth Perry, took her own life while waiting for an inspection report in January 2023. She had found out that Ofsted was going to downgrade her school from 'outstanding' to 'inadequate'. Her family later said she was devastated and humiliated by the inspection process, as well as dreading the stigma of the impending report.

Everyone agreed that the old Ofsted one-word, quaternary system (with schools deemed 'Outstanding', 'Good', 'Requires Improvement' or 'Inadequate') was a gross insult to both teachers and the ethos of education as a whole. An independent review was commissioned, which led, in November 2025, to the introduction of ... another rating system. The difference was that this one had five points of difference, not four. Schools in the UK, after considered assessment, could now only be:

- **Exceptional:** The highest possible standard.
- **Strong standard:** A very good performance level.
- **Expected standard:** The standard where most schools will likely be graded.

- **Needs attention:** A grade for those that require improvement.
- **Urgent improvement:** The lowest grade, for those that require immediate action.

This looks remarkably similar to the old system to me. Perhaps I'm not educated to the right standard.

The arguments about how (and whether) to grade schools, nursing homes, social care, hospitals and other vital public services will rumble on and on, and will be fought on precisely the same grounds discussed in the story of the supremacy of the five-star system in Chapter Two. One-word ratings are generally agreed to be too simplistic (and, after Ruth Perry's tragedy, emotionally damaging). But the whole point of having a rating system is to offer time-poor parents some simplicity. So how much simplicity is the right amount?

No one is ever quite sure. The New York mayoral election in 2025, which elevated Zohran Mamdani to celebrity as well as office, was one of the most high-profile ballots in recent years. Two types of ranking were adopted during the process. The Democratic primary used a ranked-choice system, while the final election in October was decided by voters simply picking who they liked best.[18]

As usual, both voting systems were criticised for being either too complicated – which would likely mean that less-engaged voters wouldn't bother at all – or too simplistic. But utilising two different systems did enable comparisons between the politics that they evinced. The ranked-choice primary was marked by coalition-building and general, slightly vomit-inducing niceness between candidates who were happy to cross-endorse one another.[19] The New York City general election, which used choose-one voting, saw an entirely different tone. Debates were roundly negative and dominated by discussion of which candidates were playing 'spoiler' and who should drop out – rather than those candidates' platforms and the issues facing the city.[20]

Then again, turnouts for both votes were historically high. At one level, getting people to rank out of five had worked.*

★★★★

The jury is very much still out on whether co-opting our rating obsession for the public realm might help. Behavioural economics will tell you that 'nudging' people into doing the best or the right thing is better than leaving them to do nothing,[21] and that utilising a now well-known paradigm to get more people to vote is an obvious 'nudge'.

But the jury is decided that something needs to be done. In the UK, repeated governments have looked at our first-past-the-post electoral system† and wondered if proportional representation‡ would be better. These governments tend not to do anything about it, though,

---

* It should be noted that the STAR voters have little time for the ranked-choice system (in what seems to me a very People's Front of Judea versus Judean People's Front kind of way). STAR say that ranking between one and five (where you're saying who is better versus all the others), as opposed to *rating* between one and five (where people can have the same rating; you're assessing them independently) takes us back to tactical voting and its discontents.

† During a general election, 650 constituencies across the country each hold separate contests. To become an MP, a candidate needs the largest number of votes in their constituency. This means every MP has a different level of local support. In many areas, the majority of people will not have voted for their MP. Even if millions of voters support the same party, and if they are thinly spread out across the UK, they may only get the largest number of votes in a couple of these contests, so only win a few MPs. Tens of thousands of voters supporting a different party, but who live near each other, could end up with more MPs. This means that in the UK the number of MPs a party has in Parliament rarely matches their popularity with the public.

‡ Where the total number of national votes is linked directly to the number of seats in Parliament.

because however broken the system is, it is the system that has taken them to power.

In the 2024 UK general election, for example, Labour won 63 per cent of the seats in parliament on just 33.7 per cent of the vote. That certainly seems unfair and is partly why, after just a year in office, a government with a huge electoral mandate was nonetheless very unpopular.

Everyone wants to feel like they matter and a sense of agency is vital to democratic participation.[22] It's one reason why the whole ratings culture that is the subject of this book has flourished.

A sense of agency may even be good for individual health. Let's go back to the great pet election. How did you feel when you tapped on that link and jumped into the great philosophical farm day? Hopefully, you're a little more connected to this book, a little more part of something, a little more connected to someone else – even if that person is just me (Hello) or my dog (arf). The review process is like a conversation or a proxy relationship. You don't just get noticed; you have, if not a friend, then at least someone acknowledging you exist.

It is, I think, no coincidence that the rise of the rated society has coincided with a broader societal shift into silo-ing and loneliness. Now, more than ever, in a post-pandemic world, we are all looking for something that makes us feel like we matter.

Psychologists call mattering 'the feeling of being significant to others' and they differentiate between mattering and belonging or inclusion. In his book *The Power of Mattering: How Leaders Can Create a Culture of Significance*, the psychologist Zach Mercurio talks about how one of a baby's first instincts is the 'grasp reflex', when they reach out frantically until you give them your finger or shove your nose in their face and they grab it and calm down.

'It is an automatic response initiated in part to secure our first caring relationship,' says Mercurio. 'From our first breath, our survival depends on being important to someone.'[23]

As we grow up, that instinct evolves into a basic need to be noticed, heard and valued. Studies have shown what you probably sense already – that when people feel they matter, they do well. They're more motivated, more resilient and even healthier, whereas when people feel insignificant, they languish. Disconnection, disengagement, anxiety and depression often follow.[24]

The link between mattering and politics, the subject of this chapter, was made most famously by Hannah Arendt, writing in 1951.

'What prepares men for totalitarian domination in the non-totalitarian world is the fact that loneliness, once a borderline experience usually suffered in certain marginal social conditions like old age, has become an everyday experience.'[25]

In 2023, the US surgeon general declared that loneliness was the most serious mental health crisis facing Americans. The document was entitled 'Social connection can transform our whole health and well-being',[26] which was pretty unequivocal in itself, but it was pockmarked with shards of cutting evidence in the form of fact cards promoting the importance of social connection. 'Smaller social network size is associated with an increased risk of type 2 diabetes and of other diabetic complications including heart attacks and kidney disease,' said one fact card.[27]

'Immersion in local community-based activities such as volunteering significantly increases the likelihood of unemployed individuals becoming employed,' said another.[28]

The ratings game is like joining a miniature social network in itself, a community-based activity in which you are indeed volunteering – although you're volunteering your opinion more than anything else. The pay-off is that your opinion seems to matter. At a time when society is more atomised than ever – with people living and existing alone, unnoticed – you, my friend (and forgive the Paltrow-ism) are *seen*.

My mother-in-law, who is 89, says she responds to all of the many feedback emails that pummel her inbox daily 'because it would be rude not to'. In this sense, rating and reviewing form part of an ages-old construct that you could call societal.

It may just be that having your say, no matter how, is vital to public health in both the literal and figurative sense. The health of the body politic as much as the health of the polis's bodies.

To go back to where we began in this chapter, with Walter Lippmann's quotation, by playing the ratings game, you are living in a world that you *feel* you can direct, rather than just watching it pass you by. It's illusory, but that sense of power is a straw to be clutched at in a world that people increasingly feel *is* just a slideshow operated by someone else. It feels good to have your say. And the other option – everyone feeling bad because they feel like they have no say – is much worse.

★★★★★

Chapter Seven

Who Rates the Raters?

When Uber arrived in 2012 to disrupt/ruin the frankly perfectly serviceable urban taxi system, it introduced the masses to the delights of reciprocal ratings.

You probably know the drill: when you take a ride with Uber, both the driver and the passenger give each other star ratings based on their trip experience. Drivers' and passengers' rating are taken as an average of your last 500 trips. I'm currently a 4.81.*[1]

A two-way rating system like Uber's is effectively a measure of reputation, but it's more than just a number: it plugs into human psychology and evolutionary biology at a deep level, which is one reason why there was so much concerned chatter about Uber's system when it launched.

'Riders' (passengers, to you and me) didn't – and still don't – like it:

'Why is my uber rating a 4.5 despite never being impolite and always tipping?' asked HEhehehehe6666 on Reddit (noting that he/she doesn't *ever* slam the car door, making that missing half star a heinous injustice).

---

* The average Uber customer rating worldwide is approximately 4.82 to 4.84, based on Uber's own data (see endnote), although it varies from country to country and region to region. The London average is 4.75, so actually I am really nice, thank you.

And Uber drivers don't like it either, possibly because they soon found that they could be 'decommissioned' for 30 days if their rating dropped below 4.6, just like in *Black Mirror*'s 'Nosedive' episode.[2]

'Uber's rating system punishes drivers unfairly, makes it hard for them to succeed and keeps them in the dark about how their scores are calculated,' wrote driver Jon Mendez, echoing the broader sentiment on the Uberpeople.net forum.*

Uber's system was a classic example of ratings cutting both ways. To both driver and passenger, they were neither carrot nor stick, but a sort of sharpened carrot that can still cause ample pain if jabbed purposefully in the right area. So why did Uber adopt the system in the first place?

They say that they 'use a rating system for passengers and drivers to help make sure that everyone on the Uber app treats each other with respect'.[3]

Ratings for respect – now there's an interesting one. We talked about reciprocal altruism in Chapter Two, but it's worth revisiting because its bearing on a world where everyone is rating everyone else is considerable. Reciprocal altruism – a concept in evolutionary psychology that describes cooperation between individuals with the expectation that the favour will be returned in the future – is particularly pertinent in answering the question of why humans bother co-operating at all. Why not just be an asshole and take what you want from life? The reason that most of us don't takes us back to game theory – the study of strategic decision making – and the prisoner's dilemma (the game beloved of both economists and social scientists where participants decide whether to cooperate or compete at various junctures).

---

* Where Uber drivers hang out online, when they are not driving.

We won't go too deeply into non-random clustering producing higher than expected frequencies of positive interactions, or what strategy game theorists and biologists have concluded is optimal when a prisoner has to decide whether to be loyal to his partner or betray him.[4] But the upshot, a sizeable body of evidence now tells us, is that co-operation between humans in a game involving trust flourishes when games have *an uncertain number of rounds* and *have the capacity for our reputations to precede us.*

Buying stuff from someone on Vinted, taking an Uber ride or doing anything involving a group – and the online world is an imbroglio of anonymous social groupings – is the perfect definition of a multiple round 'game' that has the means to spread news of someone being a dick built in.[5]

You may not realise (or care) that all of this is going on as you exit your Uber, step in some gum, snag your trousers on the door frame and your phone starts bugging you to rate the driver. But someone at Uber knew their evolutionary biology: the way to get people to play nice is to tell other people how nice they are.

The star rating is the means to spread that news and, by wielding it against both drivers and the driven, Uber is aiming to foment a sense of obligation and reciprocity that will lead to a cycle of all-round good behaviour. As usual, Uber doesn't do this to make the world a better place. It does it because ratings-driven reciprocal altruism is an upwards spiral that ultimately makes more profit for Uber. (It is also an attempt to eke out bad actors, because headlines like 'Uber driver jailed for killing passenger he ordered out of car'[6] or 'Uber sued after mother-of-four driver was shot in head'[7] are bad for business.)

Although, when Uber launched, there was much chatter about its reciprocal rating system, now, nearly 15 years on, very few people get excited about their Uber rating or that of their driver – or even know what each other's is. It turns out that the system of mutual respect that Uber said they were trying to foster is less important than the bald market forces of numbers of drivers available versus numbers of riders willing to pay for them.

When Uber launched, it kept its prices artificially low to build up market share. This meant everyone started using Ubers rather than traditional taxi or minicab services. Drivers also flocked to sign up, so both sides could afford to be choosy about who they picked up and let pick them up. Now that Uber has become as enshittified* as the rest of the tech world, prices have gone up, drivers have lost interest† and no one cares about their rating. They're just lucky if they can get a car or fare at all.

★

A few years ago, I discovered that I was a 'Top Critic' on the mighty review aggregator Rotten Tomatoes ('the world's most trusted and recognised source of movie and TV reviews', according to, erm, Rotten Tomatoes). Flattered as I was (this was the first and only time my teenage children have expressed any interest in what I do), I wondered how I had earned the accolade. No one had asked me.

There is a lot of mirror-flipping that goes on in star ratings and reviews, moments when someone leaves someone else a stinker and so they respond in kind. Then, suddenly, the rater – not the product or service being rated – is under scrutiny. My elevation, if that's what it was, to Top Critic left me wondering who criticises the critics? Who, if

---

* See Chapter Five, page 149.

† In 2025, Uber introduced 'Uber Priority', which enabled you to pay more in return for skipping the queue for rides. Of course, you didn't know how long the queue was or if there was one at all when you opted to try to skip it – a classic piece of enshittification/profiteering on Uber's part.

it's not too self-aggrandising for something as silly as watching telly for a living, watches the watchers? After all, in the internet age, criticism, just like taking Ubers, is a two-way rating game.

According to Rotten Tomatoes, 'established in 2008, Top Critic is a designation that elevates Tomatometer critics who excel at their craft, are well-established, influential and prolific; they are, in a sense, the cream of the crop.'[8]

Well, thank you (I think). Nice to be creamy.*

'Top Critic status,' the vaguely pompous rubric goes on, 'is granted by a set of criteria and a selection panel and cannot be applied for.'[9]

Those 'criteria' are long and waffly, but what they amount to is, 'It's up to us.' (I assume Rotten Tomatoes can remove me from Top Critic status just as easily as it granted it, which will make my children hate me just as much as they did before.)

On Metacritic, another review aggregator, it's the same. Metacritic bestows pre-eminence by curating what it assures us is a diverse group of 'highly respected critics and publications' in fields like film, television and video games. It does not have an open application process, but rather selects critics from sources it deems to have 'quality and overall stature' (and it includes most of the famous ones, like *Vanity Fair*'s Richard Lawson, the *New York Times*' James Poniewozik, the *Sun*'s Ally Ross and the *Guardian*'s Lucy Mangan). Metacritic says it then assigns different weights to individual critics and publications, based on their prestige, to create a single, weighted average score: the Metascore.

Again, this is quite grandiloquent verbiage for what appears to be a version of like-it-or-lump-it house rules.

---

* Although at the last count there are more than 3,000 Top Critics so it's hardly a Pulitzer.

'Peer evaluation and external expertise is also taken into consideration in selecting candidates for this designation,' say Rotten Tomatoes of its Top Critics. 'We have collaborated with an advisory group of critics and industry professionals chosen for their diverse backgrounds and prominence in the field in order to ensure and maintain quality for the designation.'

So, to pick the right critics, they ask some other critics. If you're a podcaster who wants to get on Rotten Tomatoes, you also need to have ... a four-star rating on Apple podcasts in order to be granted the right to give out your own stars and so on.

My accidental arrival as a Top Critic on the top critical site for my type of criticism seems to encapsulate the starry conundrum. This is that, in a world of supposedly independent reviews, star ratings are never really independent. There is always a transactional element in which you give something and you get something back. While I have been busy doling out ratings, someone else has been rating me.

★★

Counter-rating, as I'm going to call it, is central to the rating and reviews ecosystem.

The first step is to establish if the reviews are authentic. You don't want to be giving a prefect's badge to a bot. (In the next chapter, 'Star Wars', we will see that this is easier said than done.)

Then there's the issue of whether you actually want this person as a client and then reviewer in the first place. But once the rating starts, you can't stop it. Often, customers use the rating process as a form of test, a second tier of rating.

Melanie McGovern from the Better Business Bureau explains that 'we always tell our businesses if you get a bad review respond and try to make it right because a lot of times what the consumer is looking for is not that bad review: they're looking to see how the business responds, especially with our complaint process as well. "Oh my God, I got a complaint!" It's not bad, it's not terrible. It's a great way to show how you do customer service.'

It pays, too. In their paper 'Managing Complaints to Improve Customer Profitability', the Spanish academics Jesús Cambra-Fierro, Iguácel Melero Polo and F. Javier Sese concluded what you probably already intuited. 'As more and more customers become more demanding and the number of complaints increases, firms should devise new ways to manage them more effectively to obtain a positive return on their investments and enhance financial performance.'[10]

A well-handled complaint, in other words, can be more valuable than no complaint at all.

This is the start of the ratings-go-round, when a business responds to a rating or a complaint. It's all public, everyone is continuously assessing based on the available evidence and, by rating and replying, you're adding to the evidence pool.

'As a small business owner,' says Danielle Dybiec of that travel agency, Nine Muses, 'it's very important that when someone does take the time to leave a kind review, the way you respond publicly also shows the public how you interact with your clients. And it's so simple: you use some variation of, "You were a dream to work with" or "I look forward to the next time". I don't make it about me, I make it about them. I even look to that when I read larger platforms reviews. If it's a hotel, I look to see if the management make a reply. Are they at least paying attention?'

Sometimes, the management are not just paying attention, but they are rating the raters themselves. Restaurants that have suffered under the ratings system, for example, have started using their own rating system to bite back.

At Dorian, a posh restaurant in London's Notting Hill frequented by the likes of Sir David Beckham, the owner Chris D'Sylva has turned the tables on the notion of the customer being always right. He's the one who's always right.

'It's a tiered system whereby we rank how much we like the customer and the value of the customer, or the destructiveness of the customer,' D'Sylva told the *Daily Mail*, revealing a behind-the-scenes booking system used by staff at the restaurant that ranks and rates customers based on their dining etiquette. 'It's just like any sales business does.'[11]

Does any sales business do that? Uber certainly does. I looked into it for this book, as it happens, wanting to keep tabs on who was hammering me in the reviews and how I could get back to/at them. It turns out that modern customer relationship management (CRM) platforms are able to collect, analyse and act on customer feedback to assign internal 'ratings' (for instance, a customer health score or a loyalty score). Beyond satisfaction, businesses can rate customers based on metrics such as profit generated, total revenue or how quickly they pay invoices. This helps prioritise high-value clients, which is presumably why Dorian does it.

GlassTable is a fledgling private review system for fine dining and luxury hotel staff. (The name plays on the employer ratings app Glassdoor, where employees can stick it to their bosses.)[12]

GlassTable claims that once it launches fully (it's in beta at the time of writing), it will empower servers and hotel staff to provide exceptional service tailored to each guest's needs, while creating a more efficient, transparent and rewarding work environment. Its credo is similar to that of Uber.

'As the platform scales, guests will know that they are accountable for how they treat their hospitality workers, creating a powerful incentive to treat waitstaff with kindness and respect. This fosters a culture of mutual respect, where great service is rewarded, and positive interactions become the standard.'

All of this is a very polite way of saying, 'Buy our software to root out the twats.'

But just like Dorian's system, and all the other ratings systems restaurants and hotels are operating that we don't know about, GlassTable is not public. It is visibility that makes the ratings-go-round go round. In most cases, when your review is posted, it is stuck up on a pixelated billboard that shows everyone that you are active, aware and, as we saw in the last chapter, determined to have your say.

The big companies have gamified this low-level self-affirmation into a feature. Raters are continually being ranked and rated themselves. Google has its Local Guides programme, while Tripadvisor has its Top 1% programme and offers member achievement badges. On Booking.com, you can make it (if you really want) to Genius, while Yelp calls them Elites. On the restaurant site Zomato, you can graduate from a Foodie to a Big Foodie to a Connoisseur and so on.

As with so much in the tech world, this is all founded in some deep-seated psychological voodoo. Some of it you may have read about already – the dopamine hit of the Facebook 'like', 'the goal-gradient hypothesis' that sees gamers want to make it past one more level. But what's really happening when you start climbing these contrived achievement ladders is that you're being asked to keep climbing, which translates as doing more reviewing. Remember, for the review platforms, this is a numbers game and more reviews is always better. Whether it's baubles that lead to higher social recognition and status (You're not just a carping ninny, you're a LEVEL 3 LOCAL GUIDE!), or social comparisons and competitions that encourage you to review even more than that paltry Level 2 Guide, or whether it's group identification, fostering a sense of belonging and solidarity, the goal for the companies is the same: engagement. You need to rate, rate well, review more, repeat.[13]

That's why, in 2025, Trustpilot decided that 20–24 October was Write a Review Week, during which they removed invitation caps for all US businesses, both free and paid 'in order to help you collect as many authentic voices as possible'.[14]

'We're launching Write a Review Week in partnership with National Today as a cultural moment to celebrate transparency and give customers the mic,' Trustpilot declared. 'It's a great opportunity to call for more authentic voices and build customer trust.'

And also a great opportunity to rack up reviews, because 'what starts as one review quickly becomes a real growth driver,' as Trustpilot put it. Reviews equal profit; more reviews equal more mazuma.

More reviews also equal more data. Every time you click on a star rating or leave a pithy review, you are, probably unwittingly, handing over a vast trove of low-hanging ID. Look at any rating left on a site like Tripadvisor and you will see four data points: the username, the star rating, the review content and that user's other reviews.

Under the hood, however, it is very different.

'If we go internal, then we can actually find *thousands* of data points,' one staffer at a major ratings platform told me. 'They are fingerprints and you can track them.'

The amount of power left on your phone is visible to some review platforms. So are your device settings, connectivity, IP addresses and even what kind of email and/or internet provider you're using. Are you using Google? Are you using Chrome? Are you using Bing (of course you're not using Bing).

In other words, in order to have your say, you are revealing plenty about yourself.

With all this in mind, the question is, why would you bother? And the answer is that most of us don't. Perhaps the greatest barrier to rating, to borrow a phrase rarely seen in political science papers, is that people just can't be arsed. The average response rate for reviews is somewhere between 5 and 10 per cent, depending on who you ask, while website pop-up surveys fall even lower.[15]

That's why the big companies keep trying to get you to respond more. As we saw with the founding of Amazon (as well as Trustpilot's slightly desperate Write a Review Week, which sounds about exciting as World Lawn Day), the ratings economy relies on everyone rating and getting rated with furious abandon.

★★★

As for me? I'll concede that being crowned a Top Critic (once I found out) on Rotten Tomatoes gave me the minutest Jiminy of self-pride. But the reason I write reviews is simple: it's because I am paid to do it.

Yes, people want and need to be seen and valued, as we inferred in the last chapter. But, over and above the psychology, more than anything, people like to get something back.

'Do people just want to be listened to? Sure. But my view of the world is that everyone has an incentive for what they're doing. All reviews are created for some incentivisation or other.' Those are the words of a senior staffer in Amazon's Review Programme when he talked to me. (He must remain unnamed; Amazon don't like to, or don't feel the need to, engage with journalists like me. It's a feature of major review platforms that they want people to trust them, but they are remarkably untrusting themselves.)

What, then, is the incentive behind Amazon's reviews?

'At Amazon, the normal, good customer is incentivised by the fact that, one day, they might win the lottery and become a Vine customer,' the Amazonian (as Amazon calls its employees) went on. 'Because if you're in Vine, you'll never pay for a product for the rest of your life. You can get anything you want.'

Welcome to Vineland. Contribution-based recognition programmes – badges and the like – are all well and good, but relentless customer obsession has been a central part of Amazon's approach since the company's literal day one (one of the early names Jeff Bezos considered was Relentless). And Amazon's review teams know this about their customers.

'They want the free stuff,' as my Reviews staffer put it.

They get the free stuff via Vine. Amazon Vine is the grandaddy of incentivising reviews.[16] Depending on your position, incentivising is a euphemism for bribery – or possibly just the oil that keeps the fancy-free wheels of commerce a-turning. It began in 2007 when Amazon introduced an invitation-only initiative that selected top reviewers and proceeded to send them vast quantities of free stuff. All they had to do was rate and review. And then rate and review some more.

'Vine is a customer review programme that enables Amazon selling partners to put their products in the hands of Amazon reviewers that Amazon has invited to participate in Vine,' says Amazon. 'These reviewers, also known as Vine Voices, are selected for their ability to post insightful reviews on their Amazon purchases.'

For the sellers, the ones giving away all the free stuff, the Amazon Vine programme is a way to get the ratings ball rolling. They need it to sow some kudos and reap some early customer feedback for new products. These initial reviews are like the icing on the cake of market research, allowing them to understand customer expectations, identify potential product improvements and ultimately create better products.

Selection for Amazon Vine, which continues to this day in markets including the UK, Japan, Canada, South Africa and across the EU, is like being elected to some sort of cardboard-box Valhalla, where the storied heroes of the reviewing world feast for eternity. It is the apotheosis of the rate-and-review-go-round. By filing good reviews (and we'll come back to what makes a review good or not), you are giving the copious small businesses trying to get some traction for their products on Amazon a leg-up. In return, they are more than happy to send you said product.

Of course, incentivised reviews – bribery, backhanders, whatever you want to call it when there are lawyers in the room – are illegal under Amazon's policies. The Vine programme is quick to point out that its reviews are independent; in fact, you're duty bound to be scrupulously fair. That's a large part of what makes you one of the Vine-elect in the first place.*

At first, I thought the name Vine must hail from 'Grapevine', as in a means to foster word of mouth. But I now think it might come from the idea of an umbilical cord from the Amazonian mothership that feeds the entire retail foetus – reviews beget purchases beget reviews.

But for many, the word 'vine' could easily be prefixed with 'di'. An invitation to join the Vine programme is something close to a capitalist beatification. No one knows how you get selected, but when you you're touched by the hand of Bezos, you are elevated to a higher plane. Now your voice doesn't just matter, it's vital – it's got you free stuff. On the many Vine forums worldwide, you hear countless stories of people whose entire existence is predicated on Vine deliveries they didn't pay for.

'Once enrolled in Vine, Voices may request products from thousands of brands selling in the Amazon store, which are delivered to their doorsteps at no cost,' says Amazon. 'They then use the products and provide insightful reviews that reflect their honest and unbiased opinions – positive, neutral or negative.'

---

* Reviews of a product gifted by Vine appear in the same place as all the other reviews, although Amazon Vine reviews are distinguished with the magic 'Vine Customer Review of Free Product' badge for full transparency.

What makes an 'insightful' reviewer? It's hard to say. One UK-based Vine old hand told the *Guardian* in 2016 that 'it all started with Amazon contacting me after I'd posted five to ten reviews for purchases that were quite well received. I was invited to its Vine reviewing programme in 2011 and after ignoring the emails for a while (genuinely thought it was spam or a scam), I found out it was legitimate and started reviewing everything from steam cleaners to top-of-the range TVs.'[17]

Dig deep and Amazon's website offers the following guidelines on how to improve your insightfulness rating:

> 'An Insightfulness Rating can be influenced by (but not limited to) the following:
>
> *Detailed Feedback:*
> Reviews that go beyond general statements like "good product" and provide specific details about the product's features, performance, and potential issues are considered more insightful.
>
> *Helpful Suggestions:*
> Offering constructive criticism, suggesting improvements, or highlighting potential uses for the product can increase insightfulness.
>
> *Clarity and Conciseness:*
> Well-written and easy-to-understand reviews
>
> *Addressing Customer Concerns:*
> Reviews that anticipate and address common questions or concerns that potential buyers might have are considered more valuable.

Compliance with Guidelines:
Adhering to Amazon's review guidelines, including avoiding promotional language and focusing on the product itself, can also contribute to a higher insightfulness rating.'

So insightfulness is like a measure of retail enlightenment. Brilliantly, there are AI tools that now allow you to rate your own insightfulness.[18]

If you've read all that and you're thinking, *This sounds terrible, we're all going to hell in a handcart ... but FREE STUFF. How do I join?*, then there is plenty of advice as to how to write Vine-bait reviews available if you can just infiltrate the (highly clubbable) Viner-only forums.

(Viners are careful to not talk about Vine publicly, much in the same way that freemasons and people who find rat-runs or broken slot machines sense they have been presented with something wonderful and would be best served not to share it. For Vine reviewers, identifying information – names, email addresses, websites – can be seen as trying to blag free products from non-Vine sources, which Amazon can decide is grounds for removal from the programme. And that would be the worst thing ever.)

Amazon Vine exemplifies the whirling circularity of the ratings game. Initially, it gives the user a sense of great power and self-worth, as well as 30+ parcels arriving on their doorstep each day and a whole lot of cardboard to recycle. But, over time, Viners become disillusioned and the penny drops: it is a burden as well as a blessing; you have to keep reviewing to keep your insightfulness score high and stay on Vine. (The penny also drops that you have to pay tax on anything you're sent, as Amazon Vine products are considered taxable income in both the US and the UK.)

Then, suddenly, it's your own rating that you are spending a lot of time clinging on to. Can you bear to drop down a level? How will you admit to all your chums on the forums that you are no longer in

Amazon's exalted gold tier, those who get first access to the best swag? As such, these days the overarching mood among Viners prepared to talk about Vine seems to be despondency. There really is no such thing as a free lunch.

★★★★

In addition, Amazon has changed. As Chinese imports and cheap knock-offs have flooded the market via sellers like Temu* and AliExpress, the everything store has had lots more everything it needs to sell. But that means the free stuff it sends out has fallen in quality. Hence the ire of this committed Viner:

'Up until three years ago, this programme was nothing BUT good finds, name brands and quality items. Every pick was a good pick, no matter how big or small. And you usually had a few DAYS to decide if you wanted something or not. So please don't blame us for suddenly becoming "picky" after a decade or two in the programme. And guess what – the thousands of people they keep adding ALSO don't want the Temu junk, so it still sits, and thus they keep adding more. Now instead of 5,000 people getting all good items, every day, 50,000 people (or more) are fighting for ten good items. And those usually go to bots or people who pay for it.'[19]

There is now so much stuff on Amazon – the 'Temu junk' as BeachGirl6848 puts it – that it all needs reviewing if it's to have a fighting chance of selling anything at all. Amazon has even introduced its own Temu-buster, Amazon Haul, which specialises in things you didn't know you needed at prices that defy the laws of rational

* Temu is an online marketplace launched in 2022 that offers a vast range of ultra-low-cost products – from clothing to electronics – shipped directly from manufacturers in China to international customers.

economics. And so it needs more reviewers, which is why, at a conservative estimate, the Vine programme has expanded by tens of thousands in the last five years. If you're not on it yet, just start getting insightful. They'll soon rate you back and grant you a silver tier.

Other incentives are available. The Amazon Reviews staffer I spoke to showed me a handout that had come with a dog lead bought on Amazon asking for customer feedback, stating that 'for every review we receive we will donate a cup of pet food to a local shelter or rescue!'

Obviously there was no way of knowing if the seller *would* donate a cup of pet food, or which shelter they gave it to, or whether it had anything to do with the stock-shot sad doggie on the handout that had been employed to pug-tug at heart strings.

'Notice what they're actually advertising,' the Amazonian said. 'The benefit here isn't financial. They don't pay you. *You're* not going to benefit, but they're still playing on your altruism.'

There are companies on Amazon that will hand-write postcards and send them to you, requesting reviews. There are others that, if you keep buying their schlocky products and reviewing them favourably, will start sending you free stuff directly. No review is an island; rating is a process.

Yet, intriguingly, what my Amazon contact who worked on the Reviews team ended by saying was that, ultimately, they thought the review-go-round was a good thing. In their opinion, it led people to make better, more informed and therefore ultimately less instantly disposable purchases. That was why they did it for a living.

'What I like about the review stuff is it makes the world a better place. That's why I'm here now. I could have worked for Wall Street or finance and probably make ten times more money. I like working at Amazon. If I buy a crappy product for ten bucks on Amazon and it doesn't work and I have to throw it in the trash ... that's just landfill.

But using my skills, with reviews ... that's one thing I can do that can make the world a better place. Make those ten-dollar purchases better.'

Amazon is lucky to have them. I'm not sure if facilitating the great review-go-round is making the world a better place, or if it is just making Amazon and the like more money. Jeff Bezos, the man who founded a company with ratings and rater-ratings at its centre, would probably say it's both. He wasn't available for insightful comment.

★★★★★

Chapter Eight

Star Wars

Let's say you've got an exciting new product. It could be a phone case that doubles as a back scratcher or an inflatable shirt or even a brilliant new book like this one. As we've established, in order to gain any traction in the modern retail world, you need star ratings and reviews. So where do you start?

One place would be India. In 2012, Curtis Boyd, the nursing student we first met in Chapter Five, was looking to automate the review-busting business he'd begun by helping a doctor take down a fake review. Curtis is an enterprising type – and smart with it. He'd taken himself off to MIT and learned to code. Now he was looking to write some fake-review scuppering software, which was to be a free tool to find out if a business has fake positive reviews on Google Maps. (Fake negative reviews harm businesses; fake positive reviews harm consumers.) In machine-learning-land, more data equals better tools: Boyd needed a massive data set of at least half a million fake reviews to teach his software what to look for.

What he did was incredibly clever. He went to Google and typed 'Buy Google Reviews'. The very first listing hit the mark; in fact, all of the first page of results on Google were open invitations to hack Google. For the right fee, cottage industries – largely based in India, Pakistan and Bangladesh – were (and still are) happy to sell him some dodgy reviews. The going rate was about $5 a pop.

A whole dataset of reviews was another matter. You would have thought that no one would want to admit that the reviews they were selling were fake, let alone sell all of their other customers out by offering up their entire dataset for money. Nonetheless, Boyd soon found a contact who not only wanted to flog him a dataset, but who also invited him to India for the full farm-to-table experience.

So he flew to India. (When I met Boyd in New York late in 2025, he showed me a picture of him charming a snake in a very questionable captain's hat. Him, not the snake.) And so Boyd found himself in Mumbai, staying with the man who ran a warehouse full of people, mainly women, who spent all day creating fake profiles and filing reviews.

'I could not believe my eyes. This guy said he was taking home $1m a month, but that all these people who were turning out all these reviews loved it, because they were getting much more than the 20 or 30 cents a day they would normally get.'

Boyd was struck by the destitution he witnessed. 'When you see the overwhelming poverty in India with your own eyes, it's easy to understand why so many take to careers in black hat reputation management.* There's zero punishment in these countries for fake reviews and there's a huge demand for it here in the USA.'

He ended up with almost a million fake reviews, which he fed into his software to teach it what to look for when it scoured the internet for fake positive reviews. Now, using what the software learned from these and many other fake reviews, Boyd's Transparency Company can (it claims) uncover which reviews are legitimate and which aren't

---

* Black hat reputation management means the shady business of creating fake ratings and reviews as a means to burnish or harm a person or company's standing.

– on your listing and that of your competitor – 'so the playing field stays fair'.

Depending on whom you speak to, the playing field is about as fair as quicksand. Kay Dean thinks that the whole system of ratings and reviews is completely broken. Dean first became aware that there might be a problem in 2017 when she underwent a medical procedure with a highly rated doctor in the US – and it went wrong.

'I'd put a lot of blind trust into reviews up to that point,' she says. 'I mean, it's really appealing to be able to just get all this information to make your decisions. You feel like you've done your due diligence and vetted things as best you can. So when it went wrong for me, it left me suspicious and so I just started digging.'

Dean is a very good digger. She is a former federal prosecutor with a background in investigations and she prides herself on her tenacity. But what she found when she started looking into the world of fake reviews was that there really wasn't that much digging to be done. The chicanery, she told me, was happening in plain sight.

'I google-searched someone who had reviewed this [medical] practice because they had an unusual name, just an unusual name, and it landed me in Facebook groups. Facebook has since made things more private, but at the time I started snooping around, it was like, "Oh my gosh!" This guy was a member of all these Facebook groups. One thing led to another and I uncovered just this massive amount of review trading, bartering and selling, going on in plain sight, right on Facebook.'

Dean couldn't believe the extent of the review shenanigans, nor that it was all so brazen.

'I thought, *Gosh, I'm gonna go to these companies like Yelp and Google, tell them what I'm seeing.* And then really quickly I realised *they see everything* themselves. They see everything that I do. And they're not doing anything about it.'

So Dean went off and set up fakereviewwatch.com. It's the sort of website that's so full of stories about the shady side of reviews culture that it should by rights be linked to every 'Your Opinion Matters' email, to every 'Top-Rated' banner ad, to Amazon's front page. To take just one example, most recently Dean's work and analysis, using only pen and paper notes ('I'm not techie at all. I use no automation') and a dogged refusal to let go, helped root out Pakistani fraudsters who were extorting businesses for hundreds of dollars each by threatening to post fake negative reviews on Google Maps.

'The amount of review fraud that I have single-handedly uncovered is just shocking,' Dean says. 'I can find thousands of fake Google reviews on any given day.'

Google says in its statements that it removes or blocks 85 per cent of fake reviews before the public sees them. Kay Dean rates that statement a comprehensive no stars.

'Either it's not true or it's just the immensity of the problem is so massive they can't handle it. I mean they *could* handle it if they wanted to, but they profit whether reviews are real or fake.'

And there's the rub.

★

Fake reviews are not a new phenomenon; they are as old as unfake reviews. People have been astroturfing, shilling, logrolling and carpetbagging (and seemingly always coming up with cool names to describe their sharp practice) for years. In Shakespeare's *Julius Caesar*, written in 1599, Cassius and Cinna write multiple forged letters in different handwriting and leave them in places that Brutus will find them. The letters falsely claim that the citizens of Rome are worried about Caesar's ambition and want Brutus to do something about it. He does. The fakery works.

In 2017, letters came to light that showed how, in 1913, the great French novelist Marcel Proust paid 300 francs – around £900 today – for a flattering reference to *Swann's Way*, the first volume of his *À la recherche du temps perdu*, to appear on the front page of *Le Figaro* (one of France's leading dailies then and now). Proust paid a further 660 francs for another much larger summary of a glowing review by a friend of his to appear on the front page of the *Journal des Débats*.

Thus began the time-honoured process of authors doing whatever it takes to fluff their own books. The British satirical magazine *Private Eye* publishes an annual 'Christmas Log-Rolling' column in which glowing book reviews written by authors who also happen to have books out are revealed with astonishing regularity to all be back-scratching and favours for friends.*

From the earliest days of advertising in the late 19th century, customer testimonials bought brands legitimacy. Often they were handwritten, but by whose hand it was unclear; there were no regulatory bodies or fact-checkers then.

In fact, the very act of celebrity endorsement in advertising is a form of fake review. After all, if the person is being paid to say it, their opinion is instantly dubious (unless it's Victor Kiam, who liked his Remington shaver so much he bought the company).†

---

* Full disclosure: I write for *Private Eye*, including book reviews. I do not and never have compiled the 'Christmas Log-Rolling' column, nor have I had a log rolled on my behalf.

† Victor Kermit Kiam II was the perma-tanned US entrepreneur who bought the Remington razor company in 1979 after his wife bought him a selection of electric shavers (he said) and he found the Remington to be the best. He became famous through fronting a series of TV advertisements in the 1980s in which he proclaimed, 'I liked the shaver so much, I bought the company!'

More recently, as star ratings have taken over the internet, so fraudulence has flourished. You'll have heard of the bigger scandals. For instance, in 2019 the US skincare brand Sunday Riley settled with the Federal Trade Commission (FTC) after the company was accused of posting fake reviews of their products on Sephora's website for two years.[1] But fake reviews have also hurt the little people. A BBC report in 2024 told of a small chain of independent restaurants in North Wales that was being blackmailed by criminals. Unless the owners paid thousands of pounds, the fraudsters said they would flood the restaurants' online listings with fake one-star reviews.

'The weapon is fear,' owner Sam Morgan told the *You and Yours* programme on BBC Radio 4. 'Ultimately, we are a small independent business. Yes, we have got six or seven restaurants, but actually underneath it all, we are just two lads trying hard to make a living. We are not a multi-million [pound] organisation. We don't have the resources that tackle these things.'[2]

Buying or fishing for fake reviews can be as much a product of desperation as villainy. Mike Lemon is the VP of legal affairs at the National Retail Federation in the US. He told me a story of an Egyptian restaurant that opened up next to his gym in Washington, DC.

'I was able to witness the opening of this restaurant and then the slow closing of the restaurant over a period of months. I did stop in one time to get some food and was chatting with the owner – and the energy that these small business owners approach their businesses with is really inspiring. He was there with his whole heart, and probably with most of his savings, to go on this venture.'

It didn't help. The restaurant died a slow death, in Lemon's telling because it couldn't build up a head of reviews.

'When you are a new small business, the value of reviews is immeasurable. That phase is when you're most vulnerable to feeling like you need to do something to get more positive reviews. And, as the weeks go on and you're not getting the influx that you need to sustain your business, the temptation gets even stronger to find ways to create those reviews no matter what.'

Fake reviews and gaming the system have taken hold in pop culture too. I mentioned the Goodreads review-bombing scandal in the introduction to this book, but even Rotten Tomatoes, who I really like because they said I was top at something, have been caught in the crossfire.

Rotten Tomatoes was started by some undergraduates at Berkeley, but from 2011, it was owned by Warner Bros before the entertainment giant Comcast bought a 70 per cent stake in 2016. Comcast also owns NBC Universal, which makes much of the films and television that may well end up being reviewed on Rotten Tomatoes. You can see the potential for foul play. In fact, you can see the actuality as well as the potential. In 2017, for example, Rotten Tomatoes waited more than 24 hours to post a poor critics' score for the new Warner Bros film *Justice League*, breaking with the normal practice of posting right after a studio-imposed embargo is lifted. It incensed critics and fans alike, as if the parent company were protecting its useless offspring.

This kind of thing means that many people think Rotten Tomatoes might itself well be rotten. A 2023 article in *New York* magazine about the site was headlined, 'The most overrated metric in movies is erratic, reductive, and easily hacked – and yet has Hollywood in its grip.'[3]

Stories abounded of PR companies running rings round Rotten Tomatoes, sometimes paying little-known critics, who were nonetheless in the RT pool, to either review favourably or suppress bad reviews.

The film-maker Paul Schrader, for one, did not hold back on his thoughts on how reviews have been manipulated. 'The system is broken. Audiences are dumber. Normal people don't go through reviews like they used to. Rotten Tomatoes is something the studios can game. So they do.'

Over at Airbnb, the online marketplace for short- and-long-term holiday rentals, a bad review is sometimes used as a cudgel to force 'hosts', as property owners are known, to offer refunds or discounts on future stays.

Here's one disgruntled owner. 'A guest messaged and said they had to cancel for a family matter and wouldn't be able to come. I told them they needed to contact Airbnb to cancel and reminded them about our strict cancellation policy and that we don't make exceptions, but that we could offer a discount on a future stay. They replied saying they "would hate to have to leave a review that would negatively impact future guests".'

The same mildly menacing mafioso dynamic underpins the star-rating transaction on other marketplaces like eBay, Depop or Vinted. With millions of individuals buying and selling irregularly in small quantities, a person's rating is everything. There is even a setting on eBay so that you can refuse to sell to anyone with zero ratings because they have yet to build a bow wave of credibility (which, if everyone were to use it, would instantly chicken-and-egg new eBay-ers out of existence because how do you get the kudos of that first five-star transaction?).

For businesses and individuals who rely on Airbnb or eBay or Amazon for their income, this ratings game can get very serious. There's plenty of evidence that backs up what the disgruntled Airbnb-er quoted above would tell you: bad reviews can poleaxe a nascent trader. Only 9 per cent of consumers would consider engaging with a business that has a one- or two-star average rating,[4] whereas products with an average rating of 4.5 or higher are purchased four times more than those with ratings below 4.5.[5]

The statistics are just as scary for larger businesses. It's estimated that losses due to reputation damage from fake reviews totals $152bn annually in the US alone; 54 per cent of potential customers will give up on a purchase if they suspect fake reviews.[6] I know I would.

This doesn't just lose companies or individuals money, however. It's causing an erosion of trust in general. In the era of fake news, deep-fakes and tsunamis of AI slop, almost everything comes with the whiff of phoney-dom.

A Canadian professor of law, Timothy Caulfield, offers a pithy conclusion. 'The present reality: our information environment – that space where we seek, contribute to, and interact with the world's knowledge – is completely and truly f*cked.

'It is a tangle of lies, distortions, and rage-filled rants. This has created a massive paradox: we have more access to more knowledge than ever before and, at the same time, less and less certainty about the issues that matter to us.'[7]

Fake ratings and reviews cut both ways. Fake positives stuff it up for consumers, who waste their money on duff services and poor products. Fake negatives can derail small businesses. For both buyers and sellers, fakery undermines the entire system it feeds on, like an auto-immune disease: it renders the very assurance that ratings and reviews were meant to provide redundant.

'For small businesses,' says Mike Lemon, 'the trust that you establish with your customers is more valuable than anything. If you end up with 500 five-star reviews and then people show up and the food is bad, then they've lost the trust not only in your restaurant, but also in the whole review process.'

If you're a company that only exists on a covenant of trust – one like Trustpilot or Tripadvisor or Yelp – this is an existential threat. You can argue that these companies make their money from businesses paying to list, but the trust economy is a balance bike that teeters, then falls if consumers lose faith.

Governments, also, have seen that consumers want protection. It has taken a while but, throughout the 2010s, US and state regulators and attorneys general began bringing cases against companies and agencies that posted or solicited fake reviews; states like New York pursued settlements with companies engaging in astroturfing.

Research started to prove what we all suspected. A study published by the UK's Department for Business and Trade in April 2023 found that at least one in ten of all product reviews on third-party e-commerce platforms are likely to be fake – mostly fake positives that are designed to encourage consumer purchases.[8]

It all led to two major pieces of legislation. In 2024, the FTC issued rules explicitly banning fake online reviews – covering AI-generated reviews, paid or fabricated testimonials, and sales of fake reviews – and moved to enforce them.

'Fake reviews not only waste people's time and money, but also pollute the marketplace and divert business away from honest competitors,' FTC chair Lina Khan said that August. She added that the new rules will 'protect Americans from getting cheated, put businesses that unlawfully game the system on notice, and promote markets that are fair, honest, and competitive.'[9]

A year later, the UK introduced rules under its digital markets and competition framework that explicitly targeted fake reviews and deceptive fees. Businesses were made responsible for policing and removing fake reviews, but the legislation stressed that anyone posting fake online reviews about a business could be sued for defamation and for malicious falsehood by the business owners.

Justin Madders, the then minister for employment rights, competition and markets, said: 'From today consumers can confidently make purchases knowing they are protected against fake reviews.'

Do you feel confident? I'm not sure I do – although that may be because I've sat down with transparency crusaders like Curtis Boyd who told me unequivocally that, right now, if you want to pay for five-star reviews on any platform, you still can. I'm on a Bangladeshi website right now, as it happens, where a ten-pack of Facebook page reviews is $3, delivered to you via 'Manual Services' and '100% Organic Mixed Profiles', just like the review farm we first met in India.[10]

Kay Dean says that the FTC regulations lack teeth: at the end of the day, the review platforms have no real accountability for what appears on their sites.

'You have to be able to sue 'em, actually. And here [in the US], you can't because of Section 230.'*

The star wars are not over; if anything, they're just heating up.

★★

Alex Calvert gets a ping on his phone, even as he's speaking to me. 'We had another arrest in China.' The arrest was of someone who'd been posting fake reviews. Calvert is a risk manager at Amazon of 12 years' standing who has worked on all aspects of online fraud. Most recently, he's been working full time on what he calls 'the fight against fake reviews'.

'This is illegal. These are organised criminals. This is a real problem. It is not just internet troll posts online.'

Fake reviews are enough of a problem that Amazon has dedicated teams going after what they call 'bad actors' (and they don't mean hammy thesps). His job is to identify who is posting these reviews in the real world and then decide what to do about it.

---

* Section 230 of the US Communications Decency Act provides internet platforms with immunity from liability for third-party content. It's nuts.

'It can take the form of law enforcement referrals in certain jurisdictions where it is actually illegal. I think it's unlikely that most people know that it is illegal in places and that people can really go to jail for this type of issue.'

Law enforcement is one option for Calvert and his team, but it's often not a viable one. 'You can imagine that many jurisdictions around the world have bigger problems to solve,' he notes with a smile.

When Amazon don't have that option, they look to civil routes: to the courts for things like cease-and-desist letters, onto filing lawsuits and then finally moving to litigation itself.

But how do you serve papers to a click farm* in Kolkata or a task scammer† in an Iranian garage? The reality is that a major part of the problem in what these companies cast as a war is identifying who, what and where the enemy is.

Scott Commerson, a partner at US law firm Davis Wright Tremaine LLP, has been working with Amazon and suing fake review brokers since 2021. In that time, he has brought more than three dozen lawsuits.

'The fight against fake review brokers is complicated by the fact that your average fake review broker generally doesn't have a "Contact Us" page where they list their board of directors and drop in a forwarding address,' Commerson drolly told me.

---

* A click farm is an industrial-scale operation using thousands of automated scripts, software or networked smartphones to mimic real human behaviour online. The idea is to generate fake engagement – such as likes, views and comments – to manipulate social media algorithms, boost search rankings, spread disinformation or inflate brand trust.

† Task scammers prey on people looking for remote jobs by offering them simple repetitive tasks such as liking videos, optimising apps, boosting product interest or rating things.

Review brokers are just one part of the problem. They are not the ones actually creating the fake profiles and reviews; they just solicit them. The Indian investigative journalist Shadma Shaikh has mapped out the food chain in Bengaluru. She has identified what she calls 'a sophisticated underground network of product reviewers, who receive heavy discounts or sometimes even free products in exchange for leaving glowing five-star reviews on e-commerce portals such as Amazon and Flipkart.'* (And, yes, if that sounds a little like a form of underground Amazon Vine, that's exactly what it is.)

Shaikh followed Telegram and WhatsApp groups in order to identify 'marketing agencies' that employed 'mediators' between the reviewers and the product sellers. 'The mediators oversee the execution of the entire scheme. They post deals with screenshots of products and a link to the e-commerce portal. Mediators also share the nature of the deal. In most cases, a reviewer needs to purchase the product and write a positive review. The exchange value of a deal differs. Some offer a 50 per cent discount on the product, while others offer to refund the entire amount with the reviewer getting to keep the product.'[11]

In 2023, the British investigative photographer Jack Latham published a book called *Beggar's Honey* that showed what click farms, this time in Hong Kong and Vietnam, actually look like. Predictably, they are quite boring. Latham visited workshops in people's houses and in hotels and took pictures.

'It was quite strange, it almost felt like an office – I felt like I was at a young tech startup, I don't think anybody there was older than 25, and they broke down the software and how it all works and what you can do,' Latham told *Huck* magazine. 'These phones are connected to a main computer and when a client pops up, whether it's Instagram,

---

* Flipkart is Amazon's foremost e-commerce competitor in India.

Facebook or TikTok, they then enter someone's profile, which appears on every single phone simultaneously, and you press follow or like.'[12]

Sometimes, Latham said, the phones would be operated manually, but others used a newer technique called 'box farming' where serried ranks of phones, denuded of both screens and batteries, would be wired together and connected to a computer interface like some kind of hive mind.

'There was a tremendous amount of hardware ... whole walls of phones.'[13]

Those phone walls are the frontline of the star wars and it's a frontline that stretches from India to Bangladesh and on to Indonesia, the Philippines and beyond. Some countries, like China, have attempted to crack down on these kinds of operations (the China Advertising Association banned the use of click farms for commercial gain in 2020), but they continue to pop up all around the continent.[14]

It's not hard to see why: where both labour and electricity are cheap, bandwidth is copious and regulations are non-existent, no other type of farming offers greater yields. Mapping and measuring the full extent of click farming is obviously a challenge as the whole point is not to reveal where the ratings and reviews are coming from or in what volume. But, in 2013, an investigation by the Channel 4 programme *Dispatches* estimated that there were 25,000 people in the Bangladesh capital of Dhaka using computers laboriously and repetitively for hours on end to boost the profiles of specific products to order. That was more than a decade ago; since then, the use of star ratings has exploded. In 2024, 8 per cent of the reviews on Tripadvisor, or 2.5m of them, were fake.[15] Lots of fake reviews requires lots of fake reviewers.

Nonetheless, Amazon has deep pockets and lawyers have their ways. In 2024, the company's risk team identified two websites that were selling fake reviews. They didn't know who was behind them, so they filed what's known as a John Doe lawsuit – a legal case taken out against an unknown person or party whose identity is not known at the time the lawsuit begins.

'At the time we first filed this lawsuit,' Scott Commerson says, 'there really weren't many fake review lawsuits out there, so we first had to determine what claims we would bring.'

There was also the matter of where to bring them, with the internet being a global, borderless jurisdiction. The Amazon team decided to bring the claims in Washington State where Amazon is based, and to sue for violation of the consumer protection statute on the grounds that fake reviews deceive consumers.

They also sued for violation of Amazon's terms – anyone using Amazon's website, you and me included, has already signed a form agreeing not to engage in fraudulent activity – and for interference, because Amazon's contracts with sellers, as well as with any customer who reviews on Amazon's store, contain undertakings not to post fake reviews.

Civil subpoenas were issued on service providers connected to the websites. Everyone on the internet leaves a trail and payment service providers like PayPal are a good source of information. Review sharks usually give their real information to banks because 1) they have to, and 2) they want to get paid.

Once the Amazon team had gone through several rounds of what lawyers call third-party discovery (the legal process of obtaining information, documents or testimony from a person or organisation that's not a party to a lawsuit) on email service providers and domain name registrars, they ended up uncovering a network of 77 websites all connected to the same two men. They were located in Bangladesh and the John Doe suit was amended to include their names and all of their many sites.

Perhaps unsurprisingly, the two men did not show up in

Washington State to defend their case, but a default judgement was handed down and all of the domains were seized. Now, if you try to visit one, you'll get a quiet finger-wagging telling you about what happens if you offer to sell fake reviews.

At the time of writing, however, the same company – SKITSOLUTION.BD – is freely offering a service to 'Buy Google Reviews | Safe Way To Improve Your Online & Local Business'. And if you want, you can still pick up five-star Amazon verified reviews for $20.[16] Indeed, while you're at it, you might want to grab a whole suite of Skitsolution tools for what it calls reputation management; you can bag fake up votes, fake down votes, negative reviews for your competitors' websites and all manner of digital weaponry. It's like plastic surgery for your business's reputation – quick and effective and all fake. (But don't worry: Skitsolution has 45 five-star reviews of its own from various highly satisfied customers.)

In this game of whack-a-fake-review-mole, every time the black hats* wise up, so the white hats regroup and go again.

'We even have employed something in the past called a "Walk the Line",' says Alex Calvert, 'which is you engage in undercover communication, you go up to the point of making a purchase and you say "How do I pay you?" Then you say, "Sorry, that payment method didn't work." And they'll give you five other options because they think they've got you right about to purchase something. Then you take all those insights and you just take a back seat. They'll come back to you

* In internet terminology the black hats are the bad guys, hackers who use their matchless tech skills to do ill. White hats use the same skills to do good. Wikipedia says, 'The term originates from 1950s westerns, with "bad guys" (criminals) typically depicted as having worn black hats and "good guys" (heroes) wearing white ones,' although that's not quite how I remember my '50s westerns.

saying, "Hey, come back! We want to sell you X, Y and Z!" And we won't respond. But a lot of the times, you can get a ton of information out of these individuals with a very low-cost kind of undercover communication style.'

Sometimes, the review detectives will go to PayPal with a third-party subpoena – and receive hundreds and thousands of transaction records. It's one way that Amazon are able to target 'refund-after-review' schemes, in which you buy a product, write a glowing review, email the seller to show that the review is live and then they refund you the purchase price. The refund takes place off-platform, meaning neither Amazon nor the punter sees anything other than that a genuine customer has left a review.

But if PayPal shows all the refunds seeping out of the seller's account, Amazon can say those reviews were bogus. 'There can be really effective ways that you can get not just the signals intelligence,' says Calvert, 'but also cold, hard actual data back-to-back to business use.'

Francesco Moica is a fraud analyst at the hotels site Booking.com who spoke to me in New York in 2025 at a conference for the Coalition for Trusted Reviews.

'One of the challenges that we face when we work on detecting fake reviews is that reviews can be highly co-ordinated on networks of suppliers that all have different accounts – but all those accounts share the same ownership or the same management.'

Moica talks about a big case from three years earlier that marked a major turning point in how Booking.com approach detection.

'We started seeing patterns on the types of the reviews, so the same or similar reviews that were posted for different properties. These were weird behaviours not really easy to spot on a single investigation, on a single case, but they were evidence when we started combining the data together across the entire platform.'

Booking.com decided to zoom out and put all the data together into one big set. From that perspective, they realised that all of the separate cases were just one coordinated network of fakery.

One of the questionable patterns Booking.com noticed was what they call 'impossible travel'. 'Guests' who were leaving ratings and reviews seemed to be checking out in New York at 11 in the morning and then checking in again the same afternoon – on another continent. This was the case for dozens, if not hundreds of people. The same accounts connected in different ways were leaving thousands of reviews – so they knew it was the same people behind them.

They also noticed what they call 'cross-property symmetry'. 'When we see different properties that are in different parts of the world – so different markets, and they are subject to different seasonality – they mirror each other [with similar reviews and ratings all posted at the same time] which is something that shouldn't happen,' Moica said.

Booking.com analysed 750,000 reviews and found that nearly 3,000 properties were involved. Around 2 per cent of the reviews for those properties were confirmed to be fraudulent. This might not seem like much, but set against the platform average of 0.5 per cent fake reviews, it was a notable spike.

The reviews were taken down and several accounts were permanently terminated.

If you have the word 'trust' in your name, then doubt and duplicitousness is corporate cancer. Over the years, Trustpilot has been dogged with criticism for the number of fake reviews on its platform, although it is also proud of its counter-efforts. In 2023, for example, it removed 3.3 million fake reviews – just under 1 per cent of the total on the site – the majority of which were caught by automated technology.*

---

* As always, these numbers can be construed as being meaningless as they give no sense of how many fake reviews remain up. As of early 2026,

Technology, though, can be defeated by other technology. There is still a place for real human police work. As it happens, Thomas Vermeulen, director of fraud and investigations at Trustpilot, used to be a policeman.

'Fraud is evolving and it's changing every day. Right now – and this is just a snapshot – we've got platform misusers, those that are not living up to the rules that we have set as platforms. That could be including incentives in the invitations. It could be cherry-picking, so only inviting your happy customers [to leave a rating and review].'

As Vermeulen tells me, Trustpilot is under a constant virtual siege.

'You've got your review sellers. You can go on the internet and you will find them. But you've also got scam operations. They use reviews to scam people. They say, "Hey, write me reviews for this business" and then you think that you are helping the business out, but actually in fact what they do is they try to get your personal details – and then they're gonna scam you.'

There are also plenty of fake businesses that want to get themselves a foothold on services like Trustpilot in order to seem legitimate, but aren't. And then there are the so-called reputational services, which tread the line between marketing, PR and occasionally something more mendacious.

'Not all of them are bad,' says Vermeulen, 'but some are. For example, those that help you create a five-star TrustScore in a way that's not ethical, so only inviting positive customers.'

Vermeulen explains that he and his team carry out investigations into fake reviews in the same way he used to track criminals in his police days. 'You go and look online. What can I find? What kind of

---

Trustpilot hosts over 350 million active, consumer-generated reviews on its platform.

business is this? But you can't just use the internet. Not all information online is true and not all information is online.'

Trustpilot uses what they call 'shadowings' and 'infiltration'.

'That means you start to participate in those platforms, in those groups, in those circles, in those networks where things are happening that we don't want, and we gain that information.'

And, then, of course, there's the internal data. With so many users and businesses using the platform, much is generated.

'Sometimes that one business that shows up on day one, and then another business that shows up on day three and another business on day five, we can actually see, "Hold on, that is every time the same person." I look on the internet, I start looking in Companies House and I start looking on their registration documentations, and now, all of a sudden, they all have the same method.'

Vermeulen plainly loves the cat-and-mouse element of his job.

'What we see is that, after a while, they turn off. They realise we are onto them and they try another way.'

Recently, review brokers have moved from soliciting work on public platforms to using direct messaging and encrypted services like WhatsApp. The defenders on the frontline of the star wars have had to go with them.

Bobby O'Neill is the associate director of fraud and investigations at Tripadvisor. 'You have to go find those people to receive those DMs, infiltrate the spaces that they're in. It's not enough just to say like, "Oh, let's google it every now and then." You need to live in those spaces and become somebody that they would solicit those reviews from. When you do that, they're doing half the work for you and you can just follow them.'

'Fraudsters, they always adapt,' says Thomas Vermeulen. 'It's always a continuous game. You put something in and they will try to figure out a way to get round it. It's the game you see everywhere with fraud. That's what you see with credit card fraud and that's what you see with criminals on the street. But it's what I love so much about this. It's a constant game and a constant fight.'

★★★

For Tripadvisor, that fight is more of a war. Many of their network-tracking techniques were derived from working with counter-terrorism, drug trafficking and law enforcement agencies.

They break down fraud into three categories: 'boosting', which is a person sitting there writing a glowing review for a restaurant, hotel, tour etc that happens to belong to them; 'vandalism', which is as it sounds, writing a negative review trashing your competitor; and 'paid reviews', which is buying a review from a paid programme like the Bangladeshi one mentioned above.

Tripadvisor say that paid reviews represent less than 5 per cent of the fakes on the site, while vandalism is at around 2 per cent and boosting is 54 per cent.* The typical case of fraud on Tripadvisor, therefore, is somebody sitting in their own restaurant or hotel or on tour, giving themselves five stars and telling the world how great they are. (Camel tours in Morocco, I learn, are a hotbed of boosting for some reason. North African dromedaries lead to dubious commentaries.)

The fake review sleuths like Bobby O'Neill, who spoke to me in New York in 2025, see reviews as much more than just a few stars and a couple of anodyne sentences.

'A review is a transaction that has thousands of data points, one of which is the text of the review,' O'Neill says. 'But the metadata

---

* Another expert I speak to says that Tripadvisor's paid review figure is likely way too low – boosters are easier to catch; Tripadvisor just aren't collaring the paid reviews.

around those submissions is far, far more important for us. At the very least, it helps us to identify patterns of behaviour, but also to dig into specific cases.'

When it comes to boosting, a single fake review and a five-star rating is not going to help you very much. You can write a made-up submission and you might be able to get it published, but that's not going to shoot you to the top of the list in your area. And as anyone in e-commerce or SEO (search engine optimisation – essentially how to beat Google) knows, if you're not on page one, and ideally the first hit, you might as well not exist. To do that, you're going to need to write *lots* of five-star reviews. This is where device profiling comes into play: the review sites know what device you're writing on and where it is located.

'We collect tonnes of metadata about every submission that comes in,' says O'Neill, 'and we are able to simultaneously look for patterns across that. If all of these different accounts were all submitted on the same device, and each of them, the day that they wrote the review, was the first time they ever created a review, that's worth looking at, right?'

They have to be a little bit careful. In the past, Tripadvisor has flagged up multiple reviews written on the same device, only to find that it was a tour guide passing round a single iPad for comments at the end of a trip. But it might equally be somebody who is trying to boost their own ranking.

In that case, the accounts are flagged and Tripadvisor goes searching for what's called 'open-source intelligence' or OSINT. This is the kind of detective work that any of us with a laptop and a Wi-Fi connection could do (and an acronym I intend to start using regularly to make me sound like I have a secret past in the special forces). You scour people's social media profiles, examining who they are, where they've been and what they've said. It allows the review sites to cross-reference whether the person writing the great review in fact owns the property they're plugging.

What happens if you're caught? For Tripadvisor, the punishment is less often litigation than just penalties on the site. They look to hit malfeasants where it hurts, by harming the reputation they were looking to falsely inflate.

'We focus on penalties,' says O'Neill. 'So when we catch somebody and we have evidence to be able to say, "Yes indeed, this is fraud", then their ranking will drop. Through that process, there's risk and cost involved to you as an owner trying to commit fraud. Maybe you were number 20 and you were trying to get to number one ... Well, now you've ended up at number 80 or something like that, because we're not going to recommend you to our users if we have reason to believe that you have committed fraud.'

Thus your intention to puff your rating results in you lowering your rating – a gratifyingly karmic rebuke. It is, however, a rebuke that cuts deep, as evidenced by the fact that users have started to fight back. Defending your platform, or even punishing businesses that may or may not be trying to play dirty, can be risky in itself.

The first case where Tripadvisor was sued for allegedly infringing the rights of a large tourism group that owned a number of resorts was in Turkey. According to Mark Scodie, assistant general counsel at Tripadvisor, members of staff working at a number of the hotels owned by the same corporate group had been engaged in organised boosting activity – that is, leaving five-star reviews of the hotels they worked at.

'It was sophisticated enough to be blatantly obvious to the specialists on the fraud team who smelt it based on a few factors and were then able to thoroughly document and prove it, based on the kinds of technical, internal data as well as the OSINT.'

Tripadvisor's fraud team tried to communicate with the staff at the properties, saying that they knew what they were up to and asking them whether they would mind stopping doing it.

They would mind. They didn't stop. The reviews continued to appear on Tripadvisor and so the sanctions began. Eventually, the group earned themselves Tripadvisor's red card, which is an actual big red card placed on top of all their reviews saying these people are messing with their reviews.

'It's a clear label saying that there are attempts to manipulate the system.'

This is reputational death on a site like Tripadvisor and the senior management of the Turkish group didn't like it. They sued for damages and it went to court.

This wasn't what Tripadvisor wanted – not because they didn't think they could win, but because they were concerned that court proceedings would reveal their internal digital security processes.

'You can win one battle but lose the war,' they say. They didn't want the evidence of what the group had done, nor how Tripadvisor had discovered it, to be made public. That would enable future fraudsters to circumvent it.

'In this case,' Scodie concludes, 'we were able to find a procedure in the Turkish civil courts whereby our team was able to give evidence in a way that was extremely confidential – in fact the bad actors themselves weren't allowed to see our evidence.'

Tripadvisor won the case. The Turkish group is now appealing, but the broader point is that 'entities who are looking to unfortunately play the game with reviews have more opportunity than ever to pressure you back,' as Scodie says.

Even when a company suspects fake reviews, they have to be careful about crying foul. Ratings and reviews are very big business and both sides have lots to lose.

★★★★

The fraud and legal teams at all of the big companies I spoke to describe fake reviews and ratings in martial terms. It is an ongoing battle. It's star wars, an epic campaign of good against evil akin to the war on drugs or high-level counter-terror operations. You might well ask, 'Is it, though? Who cares if there are fake product reviews out there? What does it matter to me, really?'

It does matter. A 2020 survey by the British consumer magazine *Which?* found that fake reviews made shoppers more than twice as likely to choose poor-quality products.[17] So, at the very least, you are paying indirectly by being sold tat.

On another level, for bigger-ticket purchases like a holiday, commissioning a contractor for a house rebuild or finding a doctor for major surgery, choosing a poor-quality product based on false information can be ruinous. You'll recall how it was a botched surgery that led to Kay Dean's fake review crusade in the first place. Anyone who's ever had their roof or front garden done will know that a dodgy builder can empty bank accounts and not fix a thing.

Perhaps the main reason that fake reviews matter, however, is because they chip away at the whole notion of ethics in business in the first place.

'The system as it is means you have to cheat to compete,' says Dean. 'Everybody knows everything's fake. And I think maybe that's where we're headed. There's just going to be such cynicism. Nobody's going to trust anything anymore. I think we're heading down that road.'

This is the more general erosion of trust that we spoke about in Chapter Six, one that quietly undermines a rules-based order and encourages conspiracy theorising. Eventually, it leads to the kind of sudden societal shifts that, in recent years, have rendered the world more adversarial and divided. What's strange about it is that people don't seem that bothered.

If anything, when faced with fakery – in this case in the form of fake reviews and ratings – the response is apathy. Here's one example. Fakespot, an app that could analyse product reviews and spot fakes, was acquired by the search engine giant Mozilla in 2023 'to help people navigate unreliable product reviews using AI and privacy-first tech'.

It was a great idea that has since been mothballed.

'While the idea resonated, it didn't fit a model we could sustain,' says Mozilla, meaning they couldn't afford to keep paying for something that didn't bring in any money. Fakespot and its feature within Firefox known as 'Review Checker' was shut down in 2025.

Kay Dean sees the corruption of the whole reviews and ratings system as just another part of the internet's overall enshittification.

'What started as a really novel and appealing idea – the idea of crowdsourcing and getting information, especially in our society now where so often you don't know your neighbours, so you don't even know who to ask – I think it was a real boon initially. But I say that these companies, they've created a Frankenstein's monster.'

And what of the review platforms Dean attacks? She says they don't particularly care either. Engagement is engagement and they make their money from businesses paying to list and advertise on their sites, not from the people who use them as digital credence tables.

'They're just a bunch of leeches these review platforms,' says Dean. 'They're leeching off the small business community. Just beneath the surface of this review space is this whole underworld of nefarious activity underpinning everything you see.'

I always like to finish my interviews with a really stupid question, so I sign off asking Kay Dean whether she uses star ratings and reviews herself.

'Absolutely not. The purpose of my work is to convince people not to trust or use online reviews at all. Use word of mouth. Tried and true. The whole rating system only rewards cheating. Consumers and honest businesses are harmed. And the only winners are the cheaters and Google.'

★★★★★

Chapter Nine

Beyond the Stars

Star ratings are often wrong. *Zoolander* – in my view a very funny, very silly satire on the fashion industry – rates 64 per cent on the Rotten Tomatoes Tomatometer, or three stars. That is wrong. Collated opinions very often yield duff results. At the 2003 Academy Awards, *Chicago* was voted Best Picture ahead of *The Pianist* and *Gangs of New York*. That was wrong. Ditto *Downton Abbey* winning more Emmys than *Better Call Saul*. Bounty chocolate bars, a coconut concoction created by Beelzebub himself, somehow clock in at a 4.7 on Amazon. *Housekeeping* by Marilynne Robinson, one of the best recent novels I have read, is hailed by Goodreads as 'a modern classic', but receives just 3.81 stars. Think of anything you like or that has meaning in your life, and chances are that someone else thinks it's dreck.

But we have established that human beings will always look for a way to gauge choices, evaluate other people's experiences and seek out advice on each and every decision. In the previous chapter, there was plenty of evidence that the e-commerce-led ratings and review boom, which was supposed to foster trust and confidence, has instead guided us to a world of mass fakery and large-scale consumer bamboozlement. *Mistrust* is the tenor of the modern era. For star ratings and reviews, that's an existential problem.

Perhaps, on reading this book, people will realise that they are more the fuel, and less so the beneficiaries, of the star-rating system and they'll start shunning the whole thing. Perhaps Amazon Vine will shut down and the marketing departments that foment the whole ratings merry-go-round will move on to something else as a means of persuasion. What if everyone just gets fed up with ratings and reviews as a point of reference. Could the starbursts fizzle out? And if they do, what might replace them?

None of the many people I have spoken to for this book think we are entering a post-stars era just yet, but then again, when asked, many of those same people confess to just deleting all requests for star ratings and reviews because 'I don't do them'.

Some of the companies I have spoken to aren't so sure about the system by which they've been defined either. Defaqto is a British financial services agency that calls itself 'the financial product rating experts'. Defaqto say they 'bring data, technology and consumers together to help everyone make more informed choices' about important, if very dull, stuff like insurance and pensions. They have stars one-to-five splattered all over their website, and a star in their logo, but when I speak to head of product, Dipesh Sanghrajka, whose LinkedIn bio proclaims him to be a 'customer-obsessed problem solver', he says that his company's reliance on star ratings has itself become a problem to be solved.

'It's definitely something I question because now we see stars everywhere,' he says. 'Does it get the cut-through that we want it to? Does it mean what we really want it to?' Part of Sanghrajka's job is to grapple with the problem of how star ratings are now so all-pervasive that the same symbol has different meanings in different places.

Sanghrajka differentiates between stars as used for gradation versus stars used for assessment. In some categories, such as consumer goods, you are searching for the best item, the highest star rating, and that's all. In other categories – hotels or restaurants, for example – you use star ratings to define what type of hotel or restaurant you're looking at. A restaurant with two Michelin stars is not a bad restaurant, while – with my budget and means – when I go looking for a hotel I'm actively avoiding five stars because it merely signifies something vastly poncey that I can't afford.

Star ratings in that context, Sanghrajka says, are, 'just about what's right for you. With our star ratings at Defaqto, it's very much that point – the star rating represents the comprehensiveness of a policy. Let's say you've got home insurance and the fifth star represents a policy that goes as far as to cover your garden furniture. You don't have a garden? You don't need five stars. You need the four. So we're starting to think more about suitability as opposed to anything else.'

Some companies use stars to designate different tiers of products rather than suggesting low-starred products should always be avoided. Defaqto used to sell their insurance product ratings to Tesco, who would bracket the starrings alongside their Finest, Standard and Value ranges. Not everyone needed fully comprehensive cover. Some people shop the Value range. Not everyone needs five stars.

Then there are the problems with rating itself that we've identified along the way. Not only is one person's starriness not the same as another's, but ratings are very much category-specific.

'For example, for a restaurant, if you have a 4.5 rating, that's perfectly fine generally,' says Mike Lemon, VP of legal affairs at the National Retail Foundation in the US who we met earlier. 'It's better to have a hundred reviews that come out to a 4.3 than just one review that's a 5.0.'

With big-ticket items, however, we use a different internal maths.

'I know a small jeweller who said having even one not five-star rating really does impact their business,' says Lemon. 'People have different expectations. With something like an engagement ring, they're investing so much in a once-in-a-lifetime product that they really do need all fives. And so, for them, having a single lower rating is more negatively impactful than a restaurant. When you're comparing restaurants, it's like, "Anything above four point two will do for tonight."'

Stars are therefore not a consistent measure across all cultures and applications, and you could argue that this in itself undermines the point of using them to measure things at all. We've seen how ratings are open to abuse, but ratings are also self-contradictory. They're a supposedly proportionate measurement that is disproportionate.

'You actually don't want to see 100 five-star reviews because a lot of consumers are thinking "that can't be",' says Melanie McGovern of the Better Business Bureau. 'Not everybody has a great day, not everybody has a great experience. So it's always that "take it with a grain of salt" kind of thing. Having all five-star reviews is not always the best thing for a small business.'

Of course, some businesses genuinely do have all five-star reviews. Trama restaurant in Porto, for example, was recommended to me by a friend on the basis of both his visit as well as all the good reviews. It had an immaculate 364 five-star reviews at the time of my visit, many of which commented on how it is 'absolutely worth all the five-star reviews'. It had almost become a self-fulfilling prophecy (until Google Local Guide and flagrant attention-seeker Rui Carvalho gave it four stars late last year because, in his view, the food and atmosphere weren't quite up to scratch). But then, as we've seen, Google's reviews aren't verified, so really Trama's flawlessness could mean everything and nothing at all.

On the other hand, not all one-star reviews are created equal. The YouTuber Ryan Trahan has made millions of dollars making videos in which he tests out one-star rated hotels, theme parks, airlines, drive-thrus and everything in between. He follows a simple but effective formula: dig up a few of the 'craziest', 'most concerning', 'insane' and 'absurd' one-star reviews across the various categories and go check them out.

The conclusion is always the same: they're never as bad as the reviewers say. The comedy in Trahan's videos comes from the disconnect between people's judgements and Trahan's truth, which is ironic given the point of all these judgements is to establish the truth. The underlying conclusion is that a one-star review with five exclamation marks is probably not to be trusted (although nothing with more than one exclamation mark should ever be trusted).

The British comedian Frank Skinner hosts a radio show called *One Person Found This Helpful* that ploughs a similar furrow.[1] A panel game based on 'what we all sit down and do at least once a day – shop online and leave a review', *One Person Found This Helpful* separates the star rating and the review from the item under review, leaving the panel to try to guess what they were talking about. So, 'very steep and too high' turns out to be a considered one-star assessment of Ben Nevis, Britain's highest mountain. Hopefully Ben Nevis will go away and think hard about how it can do better. What both Trahan and Skinner imply is that star ratings and reviews (including the four-star restaurant review that declares that the establishment 'deserves five stars', one of my favourites) are a better source of humour than they are of meaningful advice. What, then, might be a more useful guide in a post-reviews era?

★

One option much talked about by internet soothsayers is curation. In recent years, humanity has churned out more data than the rest of human history combined, driven by the proliferation of social media, smart devices, the Internet of Things and, more recently, AI model training.[2] The winners of the future, the curators say, will be those who can prune and filter all of this choice, those who can cut through all the chaff to make sense of our super-abundant world.

In his book *The Organized Mind: Thinking Straight in the Age of Information Overload*, the neuroscientist Daniel Levitin talks about the importance of browsing and serendipity in making choices. Effectively, what he is advocating is forgetting about star ratings, getting off the internet altogether and actually having a look at whatever items you might want to choose between – going to a real bookshop to choose a book rather than relying on Amazon's algorithm to make a suggestion, or getting a personal recommendation for a builder or lawyer or doctor rather than scouring ratings and reviews. It's about taking your time.

Doing so, Levitin says, will throw up options that neither you nor the algorithm might have thought of. Browsing online, be it for products or services, 'is not the same as holding a physical object. The procedural and cognitive trade-off at stake concerns searchability (with digital files) versus the viscerally and aesthetically satisfying act of employing the kinds of visual and tactile cues our species evolved to use.'[3]

Jumping straight to ratings and aggregating reviews bypasses all of that visceral and aesthetic satisfaction. What's more, too much choice, such as that offered by an infinite shelf behemoth like Amazon, actually lessens the chances of stumbling upon something surprising.

As the 19th-century British mathematician Augustus De Morgan said of the vast libraries at the British Museum, if something is wanted, 'it can be asked for; but to be wanted, it must be known'.[4] This makes sense to me, albeit based on a sample size of one: I often find gems in the independent bookstore or at my local record shop that I would never have had recommended to me by the algorithm, just as I would sometimes avoid a book that Amazon assures me is rated five stars. My guess is that you too still find some of your favourite music and books and TV by speaking to your friends or asking for personal recommendations.

I would argue that this holds even when it comes to services. Those 'visual and tactile cues' that Levitin talks about include meeting a lawyer or builder face to face and getting a sense of what they're like. Margaret Thatcher famously identified the Soviet leader Mikhail Gorbachev as a man she could 'do business with' in December 1984, before he became general secretary, and it may be that using your gut feel for whether someone is going to be a pragmatic partner is every bit as useful as checking their star ratings.

Similarly, the growth of community-driven tips sites like Nextdoor in the UK and the Spotted By Locals online travel guides, where recommendations come from the people who actually live there, suggest that we may be moving away from the supposed wisdom of crowds and back towards local knowledge. The likes of Facebook groups/Marketplace, or the Citizen app for safety alerts, or specialised apps like Olio for sharing items, which help neighbours connect, share information and buy/sell locally – alongside a welter of community forums (and, in the UK at least, a constantly evolving hive of WhatsApp groups) – have become where people go for a form of turbo-charged, hyper-local rating system. My experience is that they can be both really annoying and really useful.

It might likewise be time for a change in how we approach choosing culture too. Cinema Reddit threads are currently awash with the term 'phantom menacing', which is when you're so excited about seeing a film that when you see it and it's distinctly underwhelming/dreadful, you try to convince yourself and everyone else that it had worth.*[5] Part of this self-delusion and consequent confusion is born, I think,

---

* In literature, I would here like to coin the term 'infinite jesting', which I will define as when you've spent months dragging yourself through an enormous novel and so you have to find ways of justifying the effort. Hence you look for positives and meaning that you might not in a smaller book.

in the ratings game. It seeds unjustified expectations and taints honest responses. A better process, surely, is to go to see the film/read the book/listen to the record and only then immerse yourself in what everyone else thought. You might find that, away from the miasma of other people's opinions, you have rather good taste on your own.

And taste is becoming a more, not less valuable distinction.

'In an era of data abundance, the thing that is scarce is taste,' said Steven Rosenbaum in his 2011 book *Curation Nation: How to Win in a World Where Consumers are Creators*. Obviously everyone likes to think they have great taste and that almost everyone else doesn't. But look at the music algorithms on Spotify or Apple Music. They have the ability to rate music based on numbers, but they can't cross genres, see why someone who loves a bit of Bach might not like all of Bach (or get the right Bach), tell the qualitative different between (for this writer) a hunk of stripped-down Detroit techno and the (mathematically similar but totally whack) EDM of DJ Sammy. Take a look at the fashion top-sellers on Amazon or the must-have items on the popular interior design site Houzz – I would argue that, for all that star ratings and reviews have helped people make better choices, they have still to replace the noble art of discernment. But then I'm a massive snob.

Star ratings are part of a search for consensus, but consensus isn't always a good thing. In *Groupthink*, Irving Janis's classic 1982 study, the author waded through everything from the escalation of the Vietnam War to the Watergate cover-up to show how 'the psychological drive for consensus at any cost suppresses disagreement and prevents the appraisal of alternatives in cohesive decision-making groups'.[6]

Then, in their 2014 book *Wiser: Getting Beyond Groupthink to Make Groups Smarter,*[7] *Nudge* co-author Cass Sunstein (who served as administrator of the White House Office of Information and Regulatory Affairs in the Obama administration) and leading decision-making scholar Reid Hastie, both of whom we met in Chapter Two, took up the anti-consensus baton, showing how more minds aren't always better.

Get a large pool of people together, Sunstein and Hastie suggest, and it often makes things worse: 'For identifiable reasons, groups often perform poorly. In some cases, groups do not merely fail to correct the errors of their members; they actually amplify those errors. In this way, the behavioral biases displayed by individuals are replicated, and often even aggravated, within groups. It's a big mistake to think that when a group's individual members are prone to error, the group as a whole will correct those errors.'[8]

In *Noise: A Flaw in Human Judgement*,[9] Daniel Kahneman, Olivier Sibony and (again) Cass Sunstein defined 'noise' (a phrase you now hear all the time in conjunction with its counterpoint 'signal') as 'undesirable variability in judgments of the same problem'. Noise, they argue, should be blocked out in order to make good choices.

You may well recognise some of the (infuriating) tendencies listed by the decision-making wonks above as features of the rating and review pages on Amazon or Google. Errors are amplified by repetition. People follow the crowd in cascades. Extreme positions rise to the top of the sample feed and the same old complaints get trotted out again and again.

Interestingly, several of the (many) books on the subject of how to avoid groupthink and promote useful curation offer suggestions, some of which might merit consideration from the big tech companies hosting all of these ratings and reviews.

One approach for better group decisions involves silencing the 'leader' so that the views of other group members can surface.[10] Another says that we should rethink rewards and incentives – all those badges and gamified hierarchies – to encourage people to reveal their own knowledge. A third suggests thoughtfully assigning roles that are aligned with people's unique strengths – that is, making that verified reviewer verifiably know what they are talking about.

But whether the review platforms choose to improve the way star ratings do or don't work depends on there being the will to improve it. As we saw in the last chapter, *they* say they are trying. Others say it's just another sell.

★★

We've mentioned Facebook groups and WhatsApp forums already, and so you know as a consumer and as an internet user that review platforms, Amazon or Walmart aren't the only places people go to look at reviews, get information and do their research. Social media is also a huge part of it. (I even managed to get someone at Tripadvisor to admit that, in spite of all the spumes of advice and ratings that surround them at their workplace, 'social media tells me what to do. A lot.')

Whatever you think about social media, you probably use it in one form or another. Realistically, these days you have to. Older readers will be well aware of the many and varied forums for local parents and neighbours that are full of people asking for and recommending local businesses – from builders, roofers and boiler repairs to new restaurants, cafés and, on my own local chat groups, literal requests to lend sugar among neighbours.*

But if you're of a certain age and you want to see how social media might replace ratings and reviews for recommendations, you also have to try and understand how most of its users use it. One way is to ask your children.

---

* As requested by the hip-hop duo Outkast in a stand-out line from the 2003 hit 'Hey Ya!'

'Social media and how it markets to people is very much a part of my life,' says Jason Perry, trust and safety lead at the digital consultancy Sutherland. 'I have a five-year-old daughter and a 14-year-old niece and they are going to be the primary targets of the kind of social marketing schemes that are going to be coming out. So this is personal and professional for me.'

There has been a shift across generations in what people want from product recommendations. Nowadays, it's not so much about rating efficacy or benefits or cold, hard value for money as it is rating vibes.

Ben Beck is assistant professor of marketing at the Marriott School of Business at Brigham Young University in Utah. He points to Aristotle's *Rhetoric*, a fourth-century BC urtext for both writers and marketeers in which Aristotle proposed that there are three modes of persuasion: logos, pathos and ethos. Logos is a logical appeal. In marketing terms, it means product information, pricing and data. Pathos is an appeal to your emotions. It's that picture of the sad-looking dog asking you to puh-lease leave a rating. Ethos, meanwhile, is an appeal to the credibility or trustworthiness of the speaker.

Gen Z is more interested in a product's ethos than anything else. The question for them is less 'Does this product fulfil my needs?' than 'What's the vibe? Does the company care?'

Nike is a good example. In 2025, the sneaker and sportswear giant decided to swap their revered slogan 'Just Do It' for something new. This was a major sea change; in marketing circles, 'Just Do It' is generally considered to be one of the greatest calls to arms of all time, one that, after its launch in 1988, took Nike's domestic market share for athletic shoes from 18 per cent to 43 per cent in ten years, with worldwide sales increasing from $877m to more than $9bn during the same period.

Why do it?

No, that was the actual new slogan: 'Why Do It?', with a question mark on the end. But they adopted it after a year in which Nike's stock had slipped around 12 per cent. (Which was why they did it.) The problem, Nike determined, was that they hadn't been able to connect with the younger generation on their ethos, in the way that challenger brands like Hoka or On or even Gymshark or Flower Mountain had. Ironically, over time 'Just Do It' had become precisely the kind of exhortation that Gen Z hated most.

'The social media generation – they don't want to be told what to do,' says Ben Beck. 'They want to kind of come to it on their own. And thus traditional marketing, where you're *telling* someone something, isn't working nearly as well as *inviting* someone to join you.'

Where do you go to ask someone to do something social, such as join you on your putative mission (and, by the way, buy your product)? You go to social media. It has become the key battleground for persuasion, in many cases replacing ratings and reviews as a sounding board.

The influence of social media has turned the normal trust paradigms on which marketing relies upside down.

'You're seeing shifts everywhere,' Beck says. 'My mom, she would spend a lot of time on Facebook, right? You're seeing the older generation also on social media and it's very much affecting the way that they do things. My mom spent so much time on Facebook that a lot of her Christmas presents for my teen boys were these gimmicky little things that were imported – helicopters that you could bounce off the walls ... and that was the Christmas present that breaks in 12 minutes. So there's very much a trust issue happening with a lot of this e-commerce that's being promoted on social media.'

The older generation, in this telling, is very quickly *losing* trust in social media, whereas the younger generation didn't have it to begin with. Gen Z's relationship with social media is you have to *earn* that trust.

It comes back to a shift in who you believe.

'When I grew up in the '80s, advertising was extremely different,' Beck says. 'We believed what we saw. We would watch the Super Bowl just to watch the ads and we didn't think about why the brands were doing this or what was the motive behind them. It was just pure entertainment and if we saw a great ad we were like, "Right, I'm gonna go buy that."'

My first job was in advertising. I was really bad it at, but it did at least teach me one thing – advertising in the round is a crock of shite: manipulative, exploitative, playing mercilessly on psychological hang-ups to create non-existent needs. Gen Z sees advertising in the same way, but only on traditional media.

'Millennials and Gen Z in particular have an entirely different attitude towards not only mainstream media, but anything that comes through an established channel,' says Jason Perry. 'To them, that means it is from "the establishment". And that is not to be trusted.'

There has been a marked attitudinal shift. As social media has become ubiquitous, with personalities being the driving factor behind engagement in many cases, those personalities can come to be perceived as a trusted source. If somebody spends a lot of internet face time with a particular source on social media, they might form what is known as a para-social relationship, where they feel that there is trust in someone they haven't met and never will. And so they are much more inclined when advertising comes through that source to accept it at face value.

TikTok is arguably the largest social media platform in the world. It is certainly the most controversial – albeit in a crowded field. That is most likely because it was born in China and is used primarily by younger people, hence old people like me don't understand it. But Mike Dabbs does. He leads the Americas e-commerce public policy team for TikTok. He's also in charge of TikTok Shop, which we'll come on to.

TikTok is not for old people. 'If you were to go in front of Congress and ask who here uses TikTok, or who here has ever bought something off of TikTok Shop, the answer would be zero,' he laughs.

TikTok, for the members of Congress or Parliament who might be reading, is a social media and entertainment platform, but it likes to think of itself as a discovery platform too. It is one where organic videos and organic livestreams are hosted. Creators post videos that show 'I can do it so you can do it'. Users can also go on to a 'live' video and speak directly to a community and a group.

Up until autumn 2023, e-commerce on TikTok largely consisted of talking about a product and then taking a consumer away from the platform to purchase that product. But then TikTok Shop launched in the US, the UK, across Europe and in some markets in Southeast Asia. This made TikTok a one-stop emporium – see a product in a video and buy it within the app, just like you would on Amazon or Etsy or any other e-commerce platform.

For the creators making videos, this turned TikTok into a potential goldmine. Some will go on to TikTok and say, 'Hey, please go visit this restaurant. I just went there. They're struggling, it's great.' Others will do a fashion and beauty tutorial, extol the virtues of some new-fangled elixir and then go, 'Look! As luck would have it, here's a button that you can click at the bottom of the video or the livestream and you can buy it right here.'

Creators can also 'affiliate', which means that, just as in old-school advertising, they get paid to promote a product or products and there's that button again to buy it. TikTok stresses that this type of transaction requires full disclosure. Users are supposed to make it clear if they're on a brand's payroll. Countless rulings by the Advertising Standards Authority in the UK suggest that social media ad promoters either don't understand what full disclosure is, or choose not to bother.

Disclosure – 'I was paid to say I like this thing' – is a nightmare for social media companies bound by advertising regulation, but it's also easily blamed on someone else because it's almost impossible to enforce.

For example, Bang Energy is an American drink that launched in 2012. By 2020, it was the fastest-growing energy drink on the market, generating nearly $300 million in sales. And yet, that year, Bang claimed to have only spent $2,200 on traditional advertising; the remainder was through influencer work.

Whether or not Bang disclosed that they were paying huge sums to influencers was a big bone of contention among both regulators and competitor advertisers. Let's say you're Johnny Influencer with his loud voice, overblown reactions to the slightest thing and Insta360 action-cam in hand. When you're necking that can of NextBigThing Cola, who's to say that you're a brand ambassador? Maybe you just happen to have it in your video.

Then, if you introduce AI to your arsenal, you can include anything in your videos. So you don't even need to get your hands on the product and a brand doesn't have to send something to you to have you be seen to be using it. Through AI, you can have it appear in your video, make it look like you're using it and enjoying it, and therefore you can 'support' a brand without ever actually having to get your hands on the product at all.

Or, potentially, you could sell a product before it even exists to generate interest. You could make an AI version of it and see if people react to it, which could determine how you go to market with that product. We are getting into the realms of Schrödinger's Product Placement here: can you be accused of advertising something that doesn't exist?

If I sound harrumphy about the lack of disclosure on social media, ignore me. I am out of touch and, anyway, young people don't care.

'What I see is that these days people have a broad understanding that social media is going to be driven by advertising,' says Jason Perry. 'In fact, there's such an acceptance of individual contributors that it's *celebrated*.'

A (thoroughly depressing) trawl through several influencer channels on social media shows that when a channel or an individual gets an ad deal, they're actually lauded as finally making it, rather than being slammed for shilling.

'It's like, "Oh, you're worth a brand deal",' Perry says. '"Congratulations! Get that bag!" and the applause emojis come out. So rather than people saying, "Oh, you're selling out", they're very excited and, in fact, they will support that brand all the more because they're also supporting that creator through that process. It is a complete paradigm shift, in my opinion.'

Indeed, the younger generation seem to trust the individual creator more than they do the platform on which they're posting. People are fickle, platform-agnostic and more likely to stick with the driver than the car.

David Sullivan is the executive director at the Digital Trust & Safety Partnership. 'I think we've seen this most with social media platforms like Twitter-cum-X and Bluesky and how people have moved between them,' he says. 'Bluesky had a kind of explosion of moderation issues where they instigated some new rules and people felt that they were being betrayed by those new rules and so they returned back to other platforms.* So when it's platform-based, I feel like it's whichever way the wind blows.'

---

* In 2023, Bluesky, a social app that had been touted as a replacement for Twitter after Elon Musk bought (and ruined) that platform in 2022, ran into

When it comes to trusting voices and personalities on social media, the younger generation will trust someone they follow 'until something happens', as Sullivan puts it. 'At which point, they will decide to withdraw that trust. And at that point, it's much harder to regain trust on an individual basis.'

In 2017, for example, model-cum-celebrity-scion Kendall Jenner made a Pepsi ad that saw her leaving a photoshoot to join a protest march. She was roundly pilloried for reducing complex social justice movements to marketing fodder and later apologised to her many followers for her crassness. Even then, the internet has a notoriously short memory. Jenner remains one of the world's highest-paid models and, in 2021, launched her own tequila brand.

Obviously, many brands and many traditional platforms are aware of this shift in credibility from ratings and reviews in the aggregate to individuals on social media. On Amazon now, sometimes when you shop an influencer will pop up showing off different products and guiding people through them. That, surely, is an acknowledgement of how influential social media has become for recommendations. It means we have come full circle, from looking to the wisdom of crowds right back to trusting in a (paid) individual, just like when the actress Lillie Langtry first appeared in an 1882 poster for Pears Soap.*[11]

---

its own moderation issues. Users discovered that Bluesky did not prevent users from using racial slurs within their handles, so the company changed its rules and quadrupled its moderation staff. This created its own backlash as some users accused the company's new rules of stifling free speech.

* In 1882, the British socialite and actress Lillie Langtry became the first celebrity to endorse a commercial product by appearing in advertising for Pears Soap. She solidified her status as a poster-girl for the brand by signing a statement declaring, 'I prefer Pears Soap to any other', which was widely used in the company's marketing campaigns. Langtry therefore holds the

The crucial thing on TikTok is that while advertisers can pay to be in videos or lives, they can't pay to elevate that video to be what TikTok staffers call 'top of the line'. Only the algorithm decides what comes out on top of a user's feed. What that means is that the sellers who pop to the top are most often not big or known brands.

'Trust is everything. The reviews that go into the seller or the product are critical on our platform and every other platform,' says TikTok's Mike Dabbs, 'really because a lot of these brands and a lot of these sellers are much, much smaller than you may see elsewhere.'

In other words, on some social media, ratings and reviews *do* remain as important, if not more important, than they ever were. What's intriguing is whether they will stay that way.

'You have to be 18 years old or older to purchase off TikTok Shop,' says Mike Dabbs. 'That is a hard and fast rule and I think that goes for most e-commerce platforms.* You also have to be 18 and older to be creating or promoting, right? The data that our CEO has used is that 32 is the average age of a purchaser. The average purchase price is somewhere between 40 to 50 dollars.'

But if Gen Z finds its trust in other ways – if it really does idolise those who are on the make and trust people essentially for being untrustworthy like Perry says – then the carapace of the ratings system collapses. There's no wisdom of crowds if you're dead set on not following the crowd.

★★★

---

dubious honour of being the first influencer. (She was also the first acknowledged mistress of the Prince of Wales, later King Edward VII.)

* It's a rule, but as usual it's neither hardly nor fastly enforced. My entire son's year nine class are all busy buying and selling on eBay, Vinted and other sites.

We've looked at the many failings of star ratings, suggested a few ways the current system might be improved and asked whether curation, localisation and social media offer an alternative. But the technology most likely to supersede star ratings is artificial intelligence. Indeed, most of the major ratings sites already feature AI appendages with silly names (Rufus, Bard, Bing, Grok ... pretty much all of the Teletubbies are covered) that promise to further ease the process of acquiring knowledge, sifting data and making sensible decisions on what to do, buy or see.

As it does in many fields, with star ratings AI promises to solve some existing problems – but a little prognostication suggests it might create many more.

Generative AI, for example, can write reviews and assign ratings on its own (that's what's 'generative' about it). Amazon and eBay already use it to summarise all the thousands of other reviews with one precis to rule them all. Meanwhile, for the reviewer – you and me – it's simple enough to ask ChatGPT or Claude to write a product review on your behalf with just a few prompts. It's therefore perfectly possible for the very 'bad actors' we met in Chapter Eight to get AI to write slews of reviews with whole fields of ratings. And, as we've seen, they already are.

Recent studies suggest that humans are more than happy to take advice from a machine. We listen to what large language models like ChatGPT say, and will change our opinions and intentions at their bidding, whether it's about our health, our mental state or what brand of T-shirt would best suit us.[12]

Moreover, when we're asked to spot AI-generated reviews, we're not very good at that either.[13] In some contexts, AI-generated reviews are even more influential and persuasive than genuine reviews.[14] Sometimes AI coughs up some really good, purple prose. By contrast, some humans are terrible writers (go ahead, insert joke here).

Finally, remember the concept of social proof: the people staring at the sky outside the subway station who are soon joined by other people because they can't *all* be staring gormlessly at nothing? Well, that works with AI too – scale multiplies impact. Even if a single AI review isn't inherently more persuasive than a human one, AI enables rapid creation and dissemination (and paraphrasing) of bags of favourable reviews. These can strongly bias consumer decisions because the weight of numbers – 10,000 reviews can't all be wrong, right? – is persuasive in itself.[15]

As always with AI and luddite catastrophism more generally, it's not all bad news. Some of it, with a little reflection, is not news at all. From the point of view of the review sites, with the sheer number of ratings and reviews being filed every second, it is impossible for humans to be rating all those ratings and reviewing all those reviews. AI can.

For the consumer, one of the benefits that have come from AI is that it saves time: you don't have to go through multiple websites and portals to look at countless disparate reviews to get a summary of your product.

The flipside is whether you believe that summary. Cognizant, an American multinational IT consulting and outsourcing company, offers review assessment to businesses. Prem James is its director of trust and safety practice.

'AI has enabled us to use advanced models and create a semantic fingerprint on every review and then that helps us to then analyze not just the language but also the motivation, the intention behind it, the authenticity behind it.'

AI allows these companies to root out the above mentioned bad actors. For small businesses or individuals like the ones we met in Chapter Five, who don't have the resources of Amazon's crack legal team (see Chapter Eight), AI levels the playing field in the fight against fake reviews.

But is AI a good actor itself or does it keep forgetting its lines and making stuff up on the spot? It's easy to talk about AI like it's the superintelligence system Skynet in *Terminator* 2, but in truth AI is a cover-all term for a raft of different technologies. Many of them have been around long before everyone got excited about large language models and being able to chat with your laptop like it's your friend.

Remember a few years ago when the big story was big data? Before the release of ChatGPT in September 2022, big data was the new trend that spawned a thousand blogs from amped up execs on LinkedIn. Like this one: 'The basic idea behind the phrase "big data" is that everything we do in our lives is (or will soon) leave a digital trace (or data), which we (and others) can use and analyze. The advances in capturing and analysing big data allow us to decode human DNA in minutes, find cures for cancer, accurately predict human behaviour, foil terrorist attacks, pinpoint marketing efforts, prevent diseases and so much more. But big data could be used for good or evil, like kryptonite, nuclear weapons and so on.'[16]

Big data was just a form of AI, a precursor that provided the fuel for the LLMs to learn from.

'From a fraud prevention perspective,' says Pascal Mafait, senior director of fraud, trust and safety at Booking.com, 'machine learning is a form of AI that has been around for ages basically and has been helping us fighting the bad guys for ages as well.'

Machine learning uses analytics to try to identify patterns across huge pools of data and it's been scanning your star ratings and reviews for nearly a decade now at companies like Booking.com, Tripadvisor, Trustpilot and eBay to determine if you're a bot or a fraud. You can judge for yourself if it's been a success, depending on what proportion of reviews and ratings you think are fake.

Generative AI does, as the name suggests, come up with coherent sequences of words on its own, but the model is trained on your ratings and reviews. It still needs your input for its output. That mega data farm in West Virginia can't tell you what it thinks of the latest Shark vacuum cleaner; it can only make a very tasty, well-seasoned word stew of what other people think. It is like one of those 1990s pictures made up of millions of other pictures that, when you zoom out, makes an even bigger picture. But it is still, at base, a – very convincing – collage.

AI, therefore, can't replace human input entirely because it needs human input to be useful. This is one reason why everyone from Amazon to Tripadvisor continues to prompt users for that next rating and review.

One keystone of useful reviews is that they be recent reviews. Stuff changes and so do people's needs and tastes.

'Think about it,' says Melanie McGovern of the Better Business Bureau. 'You find a review for a restaurant and it was five years old. With the turnover in restaurants, are you really believing a five-year-old review or are you believing the newer reviews? So that's why we always tell our businesses it's so important to get those current reviews because you might not have the same staff you had five years ago. You might not be offering the same products and services as you did.'

The danger with AI is that it gets through such vast quantities of material that the pool of human reviews is close to running out. Current LLMs are built off product reviews or other posts or generally anything that's been committed to the internet by humans. When they run out, AI begins to feed on AI-generated reviews, which were themselves agglomerations of human input. As the spin-cycle runs again and again, the human reviews that fed the whole thing in the first place are waved off into the distance. AI is then merely giving you a fresh take on some very stale produce.

Caitlin Brosseau, global head of government affairs and public policy at Tripadvisor, sums it up best. 'We need to keep the flywheel turning with authentic user experiences.'

This is something of an event horizon that is scaring the bejesus out of the whole AI (and ratings) industry: the moment when more content has been generated artificially than by real humans. When the LLMs are built more on AI-generated content than 'authentic user experiences' – when computers are effectively asking other computers for their star ratings, not you and me – there goes your credence table.

'The reality is that this is going to happen,' says Louis-Victor de Franssu, co-founder at Tremau, an advanced content moderation company for digital platforms that specialises in AI. 'It's not something we're going to be able to stop.'

De Franssu's company have looked into how you might put protections in place that would stop these different models going far and beyond where they should go in order to train themselves. He has concluded that you can't.

'With the internet and how these models are being trained and how they're scraping all over the place, I don't see how you could actually put guardrails in. You have to build them in from the get-go to make sure that the AI can auto-reject anything that is considered too far outside of what authentic behaviour would be. I think that's one of the areas that would need to be worked on.'

Note his use of the future conditional tense: it hasn't been worked on yet, which is why AI is currently careering towards that event horizon where it's summarising AI summaries of AI precis, all scraped from AI reviews of other AI reviews.

Apart from the distressing sense of spiralling towards a dystopian abyss, there are two problems with that. One is that it is just churning over old reviews – and old reviews, as we saw above, are bad reviews. The other problem is that, as LLMs start to indulge in coprophagy, they get really boring. A robot recycling loop makes HAL get worse at writing decent summaries, not better.

Good writing depends on the avoidance of cliché. AI scraping for LLMs looks for patterns and repeats them, which is a very good way of encouraging cliché. Some LLMs have taken to trying to deliberately break the way they speak by taking vast chunks of 18th-century literature, mixing all the books together and throwing it into the LLM meat grinder, just as a means to dredge up some new coinages and formulations. But take a look at what Rufus or TipsyWipsy or Grok has to say about this or that product or service. Do you trust them?

★★★★

There is a difference between AI summarising genuine reviews and AI being used as a bot to generate reviews on its own. Among the ratings sites there is some dispute about how much AI review generation is happening.

Tripadvisor's Caitlin Brosseau is dismissive. 'That's actually a myth that I like to bust. We're not seeing a massive increase in reviews being generated by AI.'

But others say that AI-generated ratings and reviews are going through the roof. A 2025 study by Pangram Labs found widespread evidence of such reviews being posted, with the majority of them getting five-star reports and receiving the 'verified purchase' stamp. Transparency Company reports an 80 per cent month-over-month increase since June 2023, while Originality.ai noted that AI-generated reviews on the US online estate agents Zillow rose from 3.63 per cent in 2019 to 23.7 per cent in 2025.[17]

We've said already that the Federal Trade Commission (FTC) in the US – and the Competitions and Markets Authority in the UK – has banned fake reviews, with their definition of fake including AI-written ones. They have the right to fine organisations in breach of their regulations. But, whether or not there really is a tsunami of

AI-generated reviews manipulating minds, the overall effect is the same: people *think* there are more AI-generated reviews out there than ever, and so trust, in the sense of people checking reviews and then making a purchase decision, is eroded.

There's a further problem: many people now use generative AI as a tool to write all sorts of things. It may well be that there are a lot of genuine 'human' reviews and ratings being handed out that *have* been written by AI – but at the behest of a human.

Tremau's Louis-Victor de Franssu is French (as his name suggests). He speaks excellent English but he knows that plenty of people don't. 'And so AI has become a user empowerment tool as well. Somebody who has difficulty writing in English or whatever language, actually using those tools enables them to give the review.'

The plus side of all of this for the consumer comes with the recent regulations. The review companies I spoke to all called them 'a nightmare' because they vary, both in demands and punishments, across different jurisdictions.

'You can have a new regulation coming out of one state that's going to say you're *not* allowed to do content moderation and the other one that says you *have* to do content moderation,' says de Franssu. 'How are you going to do it?' (Of course, one way to do it is to contact his company, but I leave that to your discretion.)

For you and I, however, these regulations mean that companies are responsible for trying, and being seen to be trying, to ensure that their product ratings and customer reviews aren't bullshit.

There is a tertiary problem, too: in using AI too much, we are abrogating responsibility to think for ourselves. This again is kryptonite to ratings and reviews, because when you check out someone else's opinion on something, there's an underlying assumption that they can think critically and independently.

Sandra Khalil is associate director at All Tech Is Human, a

non-profit advocacy for responsible tech. 'I see a major risk here in outsourcing our ability to discern, think critically, be creative [and] problem-solve to an AI. That's a huge issue we're seeing socially and it's having a really strong impact on young people as well.'

User behaviour is changing. If I want a new phone, I'll look at the reviews for a particular phone. A newer generation doesn't do that. They go to ChatGPT and, instead of asking for a particular product, they list their requirements and then ask their trustworthy robot buddy what it recommends.

This is a whole new level of faith and dependence, but as I want to be part of the future, I decide to change my user behaviour and give the new review paradigm a go.

'Hey ChatGPT,' I say, before realising that you can't speak to ChatGPT on my knackered old laptop and checking hurriedly to see if anyone saw me talking to myself.

Then I type in 'I need a new vacuum cleaner for my home office.' (If it was *really* smart, it would say 'Hey Benji, are you sure you want to keep harping on about your vacuum cleaner? I'm worried it will affect your book's star rating ...')

I get two pages of vacu-chat within two seconds and, best of all, it lands up on the Shark Rocket ultra-light stick, which is not too far away from where I reached in Chapter Two, thus giving me a self-worth and olde-world intelligence boost.

It then says, 'If you like, I can pick the top five vacuums specifically for a small home office (with real-world reviews + where to buy) – do you want me to do that?'

Why yes, I do. But when I grind down to where the 'real-world reviews' come, the sources are sites like techbloat.com and yahoo.com – no more a real-world authority on vacuum cleaning than the dump of star ratings on Walmart.com. In other words, AI puts a very convincing gloss on what is actually a fairly grubby process.

Nonetheless, 'this particular model of buying is going to increase,' says Cognizant's Prem James. 'So the opportunity is again how do we make those reviews trustworthy?'

Conveniently, Cognizant are working on a metric of trustworthiness. Plug in any AI-generated text and it will tell you how much credence the review host should grant it. It looks at not only the review rating, but also things around that review, such as who wrote it, what language they used and so on.

'AI can take decisions [to take down or flag] if it is straightforward, but if it's not it goes to a human for moderation.'

I'm not sure about using computers to come up with a measure of trustworthiness of text generated by computers. That seems to me to be inherently *untrustworthy*, which, as we've seen, runs against the whole point of reviews and ratings in the first place. But it looks as if, for the time being at least, ratings will need to be created by humans in order to be properly trusted by other humans. At least until the robot apocalypse.

One further thing. If it's going to be ChatGPT, Copilot or Apple Intelligence (okay, not Apple Intelligence – it's crap) that's giving you star ratings and reviews, what happens when you ask it to rate a product made by OpenAI, Microsoft or Apple? Many of the AI companies are enmeshed in so-called circular financing agreements with suppliers.* Might those suppliers not want a little leg-up for their multi-billion dollar investments? Would ChatGPT, for example, ever diss Nvidia microchips, when Nvidia just put $100bn into OpenAI?

---

* Circular financing refers to a web of interconnected investments where companies invest in their partners and suppliers, creating a self-reinforcing loop that some analysts worry is inflating an AI bubble. (That definition came from AI, by the way.)

Europe is ahead of the curve on these questions, as it is on big tech regulation in general. The AI Act,[18] which came into force on 1 August 2024 with the aim of being fully applicable from August 2026, requires the makers of AI systems that generate or manipulate content to admit as much. They are legally bound to disclose when content is artificially generated.

In terms of the subject of this book, the AI Act goes further: it 'forbids AI systems which deploy subliminal, manipulative, or deceptive techniques to distort behaviour and impair informed decision-making, causing significant harm'. It also – *pace* China; *pace* 'Nosedive' – outlaws social scoring, which is describes as 'evaluating or classifying individuals or groups based on social behaviour or personal traits, causing detrimental or unfavourable treatment of those people'.[19]

What I hope this chapter has shown, however, is that whatever the regulators do, AI is going to render star rating and reviews meaningless ... *unless* we plucky humans continue committing those 'authentic' (meaning actually written by someone when they're peeved about their vacuum cleaner or delighted with their new pants) experiences to print. You may not like the interminable ratings system that has taken over the world, but if we cease to engage with it, then what will replace it is likely to be much worse.

★★★★★

Chapter Ten

Rate This Book

This book is called *Rate This Book*, and so is this chapter, and so it therefore won't surprise you to learn that, in a little while, when you're done reading, I would very much like you to do just that. Yet our tour of the rated society has shown that rating everything all of the time is like chasing your tail. We've seen how star ratings took over the world and our conclusion has been that it's great for businesses looking to learn more about their customers, great for marketers and advertisers, great for scammers looking to make a fast buck, but distinctly suboptimal for you and I sat at home trying to work out what to buy, do, see or read. It is hard to get past the words of Kay Dean, who we met in Chapter Eight: 'The whole rating system only rewards cheating.'

On the other hand, we've also noted how rating things and using ratings does come with its own rewards. There are indisputably parallels between rating/reviewing and the democratic process. At base, democracy is a system of decision-making placed in the hands of the whole population, and I hope I've made a convincing argument that everyone rating everything all of the time is, at the very least, a primer for democracy itself. Everyone 'votes' or rates to register their opinion from matters as trivial as which paperclips are best to those as vital as which surgeon should remove your gall bladder. It is the will of the collective that determines the overall outcome.

No one thinks democracy is a perfect system. Winston Churchill famously described it as 'the worst form of government, except for all those other forms that have been tried from time to time',[1] and the same is true for star ratings. They have become ubiquitous because they're the best thing we've come up with, but they're also quite often awful.

As it happens, 'the best thing we've come up with so far' is also a workable summary of reasoning using what we call the scientific method, which has served the march of human enquiry quite well over the past few hundred years. Unless you believe in divine afflatus, acquiring new knowledge relies on experimentation and amassing as much data as possible against which to test hypotheses. Star ratings and reviews are a means of conducting that kind of experimentation and acquiring that data based on a wealth of observation. It's fallible, for sure, but, like the scientific method says, it's also as solid a ground for forming conclusions about the world around us as we have.

Crucially, the scientific method recognises that, when new data becomes available, a reasonable person is more than welcome to tear up those conclusions and start again – and it's that last clause that matters to the ongoing survival of the ratings system. It feeds on the new, it needs new ratings, comments, users and their insights and even bone-headed, fractal, angry, disappointed, enraged meanderings to power the process.

The danger, therefore, is that if we stop rating things, then whatever value the system has will quickly disappear. As we saw in the last chapter, AI can't replace human ratings and reviews, it can only summarise them. Until it can deliver Daniel Levitin's aesthetic hands-on experience – and it may one day, though it can't now – AI can only rate, paraphrase and regurgitate other ratings, not create them.

My contention is that we *like* rating, too. We showed in Chapter Two how the very act of assessing and deliberating is rewarding at a biological level, but we've also seen how the anonymity and segregation of our increasingly online lives leaves us craving recognition, no matter where it comes from. Rating and reviewing offers a sense of agency and intervention. Everyone likes to feel wanted.

Big tech has identified this need, codified and monetised it. As always, look closely at the language: 'Your opinion matters', 'Have your say', 'We're listening', 'Valued customer', 'You're invited to take part', 'Your feedback really helps us get better' ... The phraseology is coddling and emotive (you have worth), but the goal is engagement. The most princely commodity in the online world is your attention.

We've looked at the various incentives for you to spend your valuable time rating and reviewing, but we should recognise that the incentives for the businesses involved are, at base, financial. In the modern economy, companies make money by meeting customer needs and this is certainly one facet of amassing ratings data: it can be helpful to other customers. But those companies make much more money selling what they know about customers to other businesses for advertising. Your views matter because your viewpoint can be packaged and sold.

As for whether you should pay any heed to other people's ratings, from what we've seen, the answer appears to be a qualified 'yes' ... but only if you know that the other people are real people. And as we saw in Chapter Eight, the battle for authenticity is raging. In this regard, star ratings have become a commodity in their own right. They are a valuable, almost vital adjunct to any product and, as such, have spawned a black market with all the criminality that suggests. Like any trade in illicit goods, it risks devaluing the genuine commodity – when the general consensus becomes that *most* ratings are fake, as opposed to just *some* of them, the rating game is over. Several of the people I have spoken to for this book think we're well past that point already.

From what I have seen, the companies who rely on trustworthy reviews for their entire existence are, unsurprisingly, doing their darnedest to stamp out the frauds. Tripadvisor, Trustpilot and, to an extent, Amazon are all, I believe, committed to sustaining the star-rating system – not out of the kindness of their hearts, but because it is fundamental to their business. The motive doesn't matter: they are trying to tame the beast they have created.

This is not the case with Google. As we saw in Chapter Four, Google Reviews remains a part of Google Maps. They're nice to have for Google rather than the backbone of the company. And yet Google's is the largest star rating and user review network in the world. If ratings are to remain trusted and viable, Google Reviews need to be credible. For example, Google is not, currently, a member of the industry body the Coalition for Trusted Reviews,*[2] despite having been invited. Perhaps Google doesn't care; perhaps they're busy. But if star ratings are to continue having any value, it will require the biggest players – who've benefitted from all of us using them – to help keep them relevant.

★

Enough doom and gloom. Right now, star ratings remain the predominant tool for consumer and critical evaluation and assessment. It means that I will soon hand this book over to the mercy of the very same system I have been describing. Every author, every publisher and every marketer I have spoken to has told me that books, in particular, live or die on star ratings. It therefore seems obvious that I should

* 'A cross-industry collaboration that has come together to promote trust in reviews across the industry through sharing best practices.'

conclude by asking you, as countless others soon will, to Rate This Book. Your opinion matters.

★★★★★

Notes

Chapter One

1 Jones, Peter, 'The Ancient Greek Art of Theatre Criticism', the *Spectator*, 2 July 2022 https://www.spectator.co.uk/article/the-ancient-greek-art-of-theatre-criticism/

2 Kinsley, Zoë, *Women Writing the Home Tour, 1682–1812* (Ashgate Publishing, 2008), p.20

3 D'Alembert, Jean le Rond, *Encyclopédie*, Preface (1751–1765)

4 Kinsley, op. cit. pp.39–41

5 Glasgow University Library, MS Gen. 738

6 Starke, Mariana, Letter to John Murray, 27 January 1832, Archive of John Murray, National Library of Scotland

7 Wertenbaker, Charles, 'The Testing of M. Thuilier', the *New Yorker*, 5 June 1954

8 'AAA History, A Century of Service', https://newsroom.aaa.com/aaa-history/ and Scarborough, Linda, 'AAA and Mobil: A Traveler's Guide to the Travel Guides', the *New York Times*, 12 January 1975

9 See: www.hotelstars.eu

10 O'Brien, Edward J. (ed), *The Best Short Stories of 1915 and the Yearbook of the American Short Story* (Maynard & Company, 1915) Also https://www.gutenberg.org/cache/epub/20303/pg20303-images.html

11 Donaldson, Sarah, 'Do We Still Need Critics in This Age of Opinion?', the *Observer*, 18 November 2018

12 Rhein, Walter, 'Star Ratings are Not a Valid Way to Assess a Book and They Need to Be Retired', The Writing Cooperative, 8 May 2023. https://writingcooperative.com/star-ratings-are-not-a-valid-way-to-assess-a-book-and-they-need-to-be-retired-b817d9349203

13 Carolina V., 'Why the Star Rating System Has Lost its Meaning for Me', Santana Reads, 30 August 2020. https://santanareads.wordpress.com/2020/08/30/why-the-star-rating-system-has-lost-its-meaning-for-me/

14 Bird, Adele, 'Book Reviews – Why I Don't Like the Star Rating System', Dellybird blog, 22 January 2019. http://dellybird.co.uk/book-star-ratings/

15 'How Fast Is the World Moving Around Us?', *McKinsey Quarterly*, Q2 2020. https://www.mckinsey.com/capabilities/strategy-and-corporate-finance/our-insights/five-fifty-the-quickening

16 'Internet sales as a percentage of total retail sales (ratio) (%)', UK Office for National Statistics, https://www.ons.gov.uk/businessindustryandtrade/retailindustry/timeseries/j4mc/drsi

17 Fedewa, Dave, Holder, Chauncey, Teichner, Warren, and Wiseman, Bill, 'Five-Star Growth: Using Online Ratings to Design Better Products', McKinsey & Company, 12 August 2021. https://www.mckinsey.com/industries/consumer-packaged-goods/our-insights/five-star-growth-using-online-ratings-to-design-better-products

18 Clark, Jon, '18 Online Review Statistics Every Marketer Should Know', *Search Engine Journal*, 7 November 2024. https://www.searchenginejournal.com/online-review-statistics/329701/

19 Cavazos, Roberto, 'The Economic Cost Of Bad Actors On The Internet – Fake Online Reviews 2021', CHEQ and the University of Baltimore, p.7, https://f.hubspotusercontent00.net/hubfs/5228455/Research/Fake%20Online%20Reviews%202021.pdf

20 'Social Choice Theory', Stanford Encyclopaedia of Philosophy, 18 December 2013. https://plato.stanford.edu/entries/social-choice/

21 Bellos, Alex, *Alex's Adventures in Numberland: Dispatches from the Wonderful World of Mathematics* (Bloomsbury, 2010), Chapter One, 'A Head for Numbers' and p.56

22 Varod, Danny, 'An Overview on Giving Scores', on Thread, 'Ratings: 3 stars vs 5 stars. Why 5?', User Experience, June 2012 and modified thereafter. https://ux.stackexchange.com/questions/23002/ratings-3-stars-vs-5-stars-why-5

23 Ibid

Chapter Two

1 Iyengar, Sheena, *The Art of Choosing* (Twelve Books, 2010), p.9

2 Ibid, p.7

3 Kokis, J.V., Macpherson, R., Toplak, M.E., West, R.F., and Stanovich, K.E., 'Heuristic and analytic processing: Age trends and associations with cognitive ability and cognitive styles', *Journal of Experimental Child Psychology*, 83 (2002), pp.26–52

4 Voss, S.C., and Homzie, M.J., 'Choice as a Value', *Psychological Reports*, 26 (1970), pp.912–914

5 Iyengar, op. cit., p.9

6 Sapolsky, R., *Behave* (Vintage 2017), pp.46–51, pp.54–61

7 Brin, Sergey, and Page, Lawrence, 'The Anatomy of a Large-Scale Hypertextual Web Search Engine', *Computer Networks and ISDN Systems*, 30 (1998), pp.107–117

8 Tilly, Richard, 'Moral Standards and Business Behaviour in Nineteenth-Century Germany and Britain', in Kocka, Jürgen, and Mitchell, Allan (eds), *Bourgeois Society in Nineteenth-Century Europe* (Berg, 1993), pp.179–206

9 Mueller, John, *Capitalism, Democracy and Ralph's Pretty Good Grocery* (Princeton University Press, 1998). Both Mueller and Tilly (above) are cited in James Surowiecki's analysis of how honesty and trust underpin capitalism in Surowiecki, James, *The Wisdom of Crowds: Why the Many Are Smarter Than the Few* (Abacus, 2005), pp.121–122

10 Using Google Books' Ngram viewer: https://books.google.com/ngrams/

11 Price, Richard, 'The M&S Trusted Value Equation', February 2026. https://corporate.marksandspencer.com/newsroom/blog/ms-trusted-value-equation

12 Lacy, Lucy, 'Price Matching Ain't What It Used to Be', *Adweek*, 18 July 2018. https://www.adweek.com/commerce/price-matching-aint-what-it-used-to-be

13 Bhaskar, Michael, *Curation: The Power of Selection in a World of Excess* (Piatkus, 2017), p.260, citing Jones, Graham, *Clickology: What Works in Online Shopping and How Your Business Can Use Consumer Psychology to Succeed* (John Murray, 2014)

14 'Nielsen Trust in Advertising Report 2021', Nielsen. https://www.nielsen.com/wp-content/uploads/sites/3/2021/11/2021-Nielsen-Trust-In-Advertising-Sell-Sheet.pdf

15 'Survey: The Ever-Growing Power of Reviews (2023 Edition)', Power Reviews, 2023. https://www.powerreviews.com/power-of-reviews-2023/

16 'Survey: The Ever-Growing Power of Reviews', Power Reviews, 2021. https://www.powerreviews.com/power-of-reviews-survey-2021/

17 Qiu, Keda, and Zhang, Liyi, 'How online reviews affect purchase intention: A meta-analysis across contextual and cultural factors', *Data and Information Management*, 8 (2024)

18 Bigne, E., Ruiz, C., Perez-Cabañero, C., and Cuenca, A., 'Are Customer Star Ratings and Sentiments Aligned? A Deep Learning Study of the Customer Service Experience in Tourism Destinations', *Service Business Journal*, 17 (2023). https://link.springer.com/article/10.1007/s11628-023-00524-0

19 Milgram, S., Bickman, L., and Berkowitz, L., 'Note on the Drawing Power of Crowds of Different Size', *Journal of Personality and Social Psychology*, 13(2) (1969), pp.79–82

20 Cialdini, Robert B., *Influence* (HarperCollins, 2007), pp.114–119, p.285 and Surowiecki, op. cit., p.43

21 'Seeing Stars: Consumers Think Products With Star Ratings are Better Than They Really Are', UBC Sauder School of Business, 8 April 2025. https://www.sauder.ubc.ca/news/insights/seeing-stars-consumers-think-products-star-ratings-are-better-they-really-are

22 Sirwani, D., Kumar, S., and Thomas, M., 'Overestimating Stars, Underestimating Numbers: The Hidden Impact of Rating Formats', *Journal of Marketing Research*, 62(5) (2025). https://journals.sagepub.com/doi/10.1177/00222437251322425

23 Surowiecki, op. cit. p.xvii

24 Boyd, Robert, and Richerson, Peter J., 'Norms and Bounded Rationality', Unpublished manuscript prepared for the Dahlem Workshop on Bounded Rationality: The Adaptive Toolbox (May 1999). https://www.researchgate.net/publication/2313089_Norms_and_Bounded_Rationality

25 Henrich, J., and Boyd, R., 'The Evolution of Conformist Transmission and the Emergence of Between-Group Differences', *Evolution and Human Behavior*, 19 (1998), pp.215–242

26 Howes, A., and Lelis, S., 'Informing Decisions: How People Use Online Rating Information to Make Choices', *Proceedings of the International Conference on Human Factors in Computing Systems*, CHI 2011, Vancouver, BC, Canada, 7–12 (2011)

27 Simon, Herbert A., *Administrative Behavior: A Study of Decision-Making Processes in Administrative Organizations*, 4th revised edition (Free Press, 1997)

28 Wilson, T.D., and Schooler, J.W., 'Thinking Too Much: Introspection Can Reduce the Quality of Preferences and Decisions', *Journal of Personality and Social Psychology*, 60 (1991), pp.181–192

29 Howes and Lelis, op. cit.

30 Pooja, K., and Upadhyaya, Pallavi, 'Does Negative Online Review Matter? An Investigation of Travel Consumers', *International Journal of Consumer Studies*, 49(2) (2025)

31 'Survey: The Ever-Growing Power of Reviews' (2023 Edition), Power Reviews, 2023. https://www.powerreviews.com/power-of-reviews-2023/

32 Kahneman, D., and Tversky, A., 'Choices, Values, and Frames', *American Psychologist*, 39(4) (1984), pp.341–350

33 Baumeister, R.F., Bratslavsky, E., Finkenauer, C., and Vohs, K.D., 'Bad Is Stronger Than Good', *Review of General Psychology*, 5(4) (2001), pp.323–370

34 Ogilvy, David, *Confessions of an Advertising Man* (Athenaeum, 1963)

35 'Why Do Consumers Provide Product Ratings and Reviews?', Power Reviews Insights Survey, 2020. https://www.powerreviews.com/why-do-consumers-provide-product-ratings-reviews/

36 'How Online Reviews Influence Sales', Spiegel Research Center, 2018. https://spiegel.medill.northwestern.edu/how-online-reviews-influence-sales/

37 'Amazon Hits Chinese Sellers With Crackdown on Fake Reviews', Bloomberg UK, 18 August 2021. https://www.bloomberg.com/news/articles/2021-08-18/amazon-amzn-cracks-down-on-fake-reviews-hitting-chinese-retailers?embedded-checkout=true

38 Bhaskar, op. cit. p.230, describes curation and selection as ethical acts.

39 Samuels, M., 'Six ways to continuously improve your products, according to business leaders', ZDNet Business, May 2025. https://www.zdnet.com/article/6-ways-to-continuously-improve-your-products-according-to-business-leaders/

40 'Why Shoppers Leave Reviews: And How to Get More Reviews', Yotpo blog, 3 August 2025. Also PowerReviews Consumer Research, 'What Motivates Shoppers to Write Reviews in 2023?'. https://www.powerreviews.com/motivate-shoppers-write-reviews

41 Wu, L., Mattila, A., and Hanks, L., 'The Impact of Power on Service Customers' Willingness to Post Online Reviews', *Journal of Service Research*, 19(2) (2016)

42 Nemeth, C., 'Interactions Between Jurors as a Function of Majority vs. Unanimity Decision Rules', *Journal of Applied Psychology*, 7(1) (1977) cited in Nemeth, Charlan Jeanne, *In Defense of Troublemakers: The Power of Dissent in Life and Business* (Basic Books, 2018)

43 Sunstein, Cass R., *Why Societies Need Dissent* (Harvard University Press, 2005), p.14

44 Ibid, pp.14–15

45 Surowiecki, op. cit., p.187

46 Sunstein, Cass R., and Hastie, Reid, 'Making Dumb Groups Smarter', *Harvard Business Review*, December 2014

47 Surowiecki, op. cit., p.188

48 'Why Do Consumers Provide Product Ratings & Reviews?', Power Reviews Insights Survey, 2020

49 Brown, J., Collins, R., and Schmidt, G., 'Self-esteem and direct versus indirect forms of self-enhancement', *Journal of Personality and Social Psychology*, 55(3) (1988), pp.445–453

50 'How Online Reviews Influence Sales', Spiegel Research Center, ed. Collinger et al, 2018. https://spiegel.medill.northwestern.edu/wp-content/uploads/sites/2/2021/04/Spiegel_Online-Review_eBook_Jun2017_FINAL.pdf

51 Posner, Richard, *Law, Pragmatism, and Democracy* (Harvard University Press, 2005), p.206

52 Armstrong, J. Scott, 'The Seer-Sucker Theory: The Value of Experts in Forecasting', *Technology Review*, 82(7) (1980), pp.16–24

53 Bhaskar, op. cit, p.231

54 Aristotle, *Politics*, Book III, Chapter 11

Chapter Three

1 Nash, R., 'You'll need to be really memorable to get five stars', *The Times*, 14 March 2025

2 'The Numbers – Movies Released in 1989'. https://www.the-numbers.com/movies/year/1989

3 Albanese, Andrew, and Milliot, Jim, 'Self-Publishing's Output and Influence Continue to Grow', *Publisher's Weekly*, 8 November 2024

4 Sirwani, D., Kumar, S., and Thomas, M., 'Overestimating Stars, Underestimating Numbers: The Hidden Impact of Rating Formats', *Journal of Marketing Research*, 62(5) (2025)

5 'Seeing stars: consumers think products with star ratings are better than they really are', Insights at UBC Sauder, 8 April 2024. https://www.sauder.ubc.ca/news/insights/seeing-stars-consumers-think-products-star-ratings-are-better-they-really-are

6 Rajaraman, S., 'Five Stars Dominate Ratings', YouTube Official Blog, 22 September 2009. https://blog.youtube/news-and-events/five-stars-dominate-ratings/

7 Bennett, Steve, 'Isn't it time the stars went out?' Chortle, 26 August 2014. https://www.chortle.co.uk/correspondents/2014/08/26/20814/isnt_it_time_the_stars_went_out

8 Cohen, Alexander, 'Curtain Call', *The Fence*, 22 April 2023. https://www.the-fence.com/michael-billington-interview/

9 Bradshaw, Peter, 'Giving a review zero stars sets a dreadful precedent. But here are the one-star shockers I'd downgrade', the *Guardian*, 6 November 2025

10 *The Rest is Entertainment* podcast, 8 May 2025

11 Mendelsohn, D., 'A Critic's Manifesto', the *New Yorker*, 28 August 2012

12 *Private Eye*, Issue 1368, 13 June 2014, p.7

13 Rampell, Catherine, 'Reviewing the Movies: Audiences vs. Critics', the *New York Times*, 14 August 2013

14 *The Rest is Entertainment*, 8 May 2025

15 Kreizman, Maris, 'Let's Rescue Book Lovers From This Online Hellscape', the *New York Times*, 24 December 2023

16 Reviews for *LA Baby* by Tim Key, https://www.amazon.co.uk/product-reviews/1916222692/ref=cm_cr_dp_d_show_all_btm?ie=UTF8&reviewerType=all_reviews

17 Carey, John, *What Good Are the Arts?* (Faber, 2006), p.49

18 Scott, A.O., 'A Times Arts Critic Reviews His Own Role', the *New York Times*, 11 December 2018

Chapter Four

1 Stone, Brad, *The Everything Store, Jeff Bezos and the Age of Amazon* (Little, Brown, 2014), p.183

2 Ignatius, Adi, 'Jeff Bezos on Leading for the Long-Term at Amazon,' HBR IdeaCast (blog), *Harvard Business Review*, 3 January 2013. http://blogs.hbr.org/ideacast/2013/01/jeff-bezos-on-leading-for-the.html

3 Cialdini, Robert B., *Influence* (HarperCollins, 2007), p.234

4 Stone, op. cit. p.227

5 'Everything you need to know about Amazon reviews', Amazon Staff, 24 July 2025. https://www.aboutamazon.com/news/retail/amazon-customer-reviews-star-ratings

6 'Create a Netflix campaign on Amazon DSP'. https://advertising.amazon.co.uk/help/GDDD3R6XBHFV8FGD

7 Mahoney, Sarah, 'Wal-Mart Adds Customer Reviews to Web Site', *Marketing Daily*, 20 July 2007

8 'Walmart Reviews vs Testimonials Survey, First Party Data', Walmart Spark Community, July 2023

9 Ibid

10 Bezos, Jeff, Commencement speech at Princeton University, 30 May 2010

11 *14 Minutes of SaaS* podcast, 'E120 – Trustpilot CEO Founder Peter Holten Mühlmann – 1 of 3 – From the Basement', 10 August 2020, Transcript here: https://www.appselekt.com/podcasts/e120-trustpilot-ceo-founder-peter-holten-muhlmann-1-of-3-from-the-basement

12 'In conversation with Trustpilot founder and CEO Peter Holten Mühlmann', *Silicon Valley Bank* online interview, 2022. https://www.svb.com/globalassets/library/uploadedfiles/svb_trustpilot_qa.pdf

13 Zink, Jeppe, 'Trustpilot and Northzone: a 150x success story 10 years in the making', *Northzone News*, 24 March 2021

14 Ackerman, Naomi, 'Trustpilot CEO Peter Holten Mühlmann: "Everybody suspects everyone is an a***hole – but people are nice"', the *Standard*, 19 January 2022

15 Feefo.com frontpage. https://business.feefo.com

16 'Which Brands Are Most Trusted in the US in 2025?', Marketing Charts, 15 July 2025. https://www.marketingcharts.com/industries/cpg-and-fmcg-236090

17 Wright, Robert, *The Moral Animal* (Abacus, 1996), p.202

18 57.5% in fact. See Vaghasia, Krunal, '20 Surprising Google review statistics (New 2026 data)', *Wiser Reviews*, 1 January 2026. https://wisernotify.com/blog/google-review-stats/

19 Ibid

20 Ibid

21 Hackworthy, Kate, 'British Chip Butty Recipe', 2 June 2021. https://veggiedesserts.com/chip-butty/

22 Google Maps User Generated Content Policy. https://support.google.com/contributionpolicy/answer/7422880

23 Johnston, Matthew, 'How Google (Alphabet) Makes Money: Advertising and Cloud', Investopedia, 5 January 2025. https://www.investopedia.com/articles/investing/020515/business-google.asp

24 Storer, Rhi, 'Head scores victory as Google agrees to delete school reviews', *Schools Week*, 4 February 2026

25 Ibid

Chapter Five

1 Researchers have shown that perceived physical and vocal attractiveness (or their opposite) lead to bias in judgement. See for example Surawski, Melissa K., and Ossoff, Elizabeth P., 'The effects of physical and vocal attractiveness on impression formation of politicians', *Current Psychology*, 25 (2006). Or Verhulst, Brad, Lodge, Milton, and Lavine, Howard, 'The Attractiveness Halo: Why Some Candidates are Perceived More Favorably than Others', *Journal of Nonverbal Behavior*, 34 (2010)

2 'Man to pay £25,000 damages over negative TrustPilot review', BBC News, 9 February 2021. https://www.bbc.co.uk/news/technology-55981600

3 AVVO Review Removal Service, Maximatic Media. https://maximatic media.com/avvo-review-removal-service

4 Varel, D., 'How and Why Colleges Should Reform Student Evaluations' Academe Blog, 2 February 2022

5 'Rate My Professors: Inaccurate, Unreliable, and Sexist', The Professor blog, Medium, 7 October 2020. https://the-professor.medium.com/rate-my-professors-inaccurate-unreliable-sexist-and-probably-racist-81a2f06047cd

6 Mackay, Brad, 'RateMDs.com nets ire of Canadian physicians', *Canadian Medical Association Journal* , 176(6) (2007). https://pmc.ncbi.nlm.nih.gov/articles/PMC1808521/

7 Ibid

8 Crowe, Kelly, 'Who's rating doctors on RateMDs? The invisible hand of "reputation management"', CBC News, 27 October 2018. https://www.cbc.ca/news/health/ratemds-privacy-reputation-management-1.4880831

9 Removify homepage. 'Fast, Effective Removal Of RateMDs Reviews'. https://removify.com/content-removal/ratemds/

10 See Crowe, Kelly, op. cit.

11 Doctify homepage. https://www.doctify.com/uk/about/doctify

12 'The Circle is a reality competition for the social media age'. https://studio lambert.com/project/the-circle/

13 Battle of the Bods. https://gameshows.fandom.com/wiki/Battle_of_the_Bods

14 Ough, Tom, 'The Circle, episode 1, review: mendacity and vanity make this a maelstrom of species shame,' the *Telegraph*, 18 September 2018

15 From Twitter user Clown in the Pantry (@talkwordy), 30 September 2015

16 Moss, Chris, 'Peeple: The cult of ratings is destroying us, one star at a time', the *Telegraph*, 2 October 2015

17 Ito, T.A., and Urland, G.R., 'Race and Gender on the Brain: Electrocortical Measures of Attention to the Race and Gender of Multiple Categorizable Individuals,' *Journal of Personality and Social Psychology*, 85 (20023): 616. Cited in Sapolsky, R., *Behave* (Vintage, 2017), p.388

18 'Powering Tinder – The Method Behind Our Matching'. https://www.help.tinder.com/hc/en-us/articles/7606685697037-Powering-Tinder-The-Method-Behind-Our-Matching

19 Tiffany, Kaitlyn, 'The Tinder Algorithm Explained,' *Vox*, 18 March 2019

20 'Powering Tinder – The Method Behind Our Matching', op. cit.

21 'What was that early 2000s photo ranking website?' Reddit r/nostalgia. https://www.reddit.com/r/nostalgia/comments/1cbnbko/what_was_that_early_2000s_photo_ranking_website/

22 Interview with *Time* Magazine, December 2006

23 Elliott, Larry, 'George Soros: China is using tech advances to repress its people', the *Guardian*, 24 January 2019

24 Lauer, Josh, *Creditworthy: A History of Consumer Surveillance and Financial Identity in America* (Columbia University Press, 2017)

25 Yu, Haiqing, 'China's social credit system as method', East Asia Forum, 22 March 2023. https://eastasiaforum.org/2023/03/22/the-social-credit-system-as-method/?utm_source=chatgpt.com

26 'Planning Outline for the Construction of a Social Credit System (2014–2020)', China Copyright and Media, 14 June 2014. https://chinacopyrightandmedia.wordpress.com/2014/06/14/planning-outline-for-the-construction-of-a-social-credit-system-2014-2020/

27 Trishala S and Kushala Agrawal, 'China's Social Credit System', *Organisation for Research on China and Asia*, 30 July 2024. https://orcasia.org/chinas-social-credit-system

28 See 'Planning Outline for the Construction of a Social Credit System (2014–2020)', *Open Research Repository* available at: https://openresearch-repository.anu.edu.au/items/2b5f1adb-6d4f-49c8-bbcb-04217523358a; Schaefer, Kendra, 'China's Corporate Social Credit System' (Congress Research Service report), *Trivium*, 16 November 2020, Trivium China.

https://www.uscc.gov/sites/default/files/2020-12/Chinas_Corporate_Social_Credit_System.pdf; and Kaiwei, Zang, and Wenxing, Zhong, 'China unveils guideline to improve social credit system', *People's Daily Online*, 1 April 2025. https://en.people.cn/n3/2025/0401/c90000-20296397.html

Chapter Six

1 Lippmann, Walter, *The Phantom Public* (Transaction Publishers, 1925)

2 Tisdall, Simon, 'France is not alone in its crisis of political faith – belief in a democratic world is vanishing', the *Guardian*, 12 October 2025

3 'Cross-Tabs: September 2025 Times/Siena National Poll of Registered Voters', the *New York Times*, 29 September 2025

4 Keate, Noah, 'Majority of UK's young adults "concerned" about democracy, new poll says', *Politico*, 31 March 2025

5 'Big decline in belief public services are listening, as Reform UK becomes magnet for those who feel powerless, study finds', Ipsos, 16 September 2025. https://www.ipsos.com/en-uk/big-decline-belief-public-services-are-listening-reform-uk-becomes-magnet-those-who-feel-powerless

6 See: https://www.starvoting.org

7 BTernaryTau, 'Why I like STAR voting: the 5-star ballot', Github, 6 June 2021. https://bternarytau.github.io/2021/06/06/why-i-like-star-voting-the-5-star-ballot

8 Ackerman, Bruce, and Fishkin, James, *Deliberation Day* (Yale University Press, 2004), p.5

9 Friedman, J. in his Introduction to *Political Knowledge* (Routledge, 2012), p.1

10 Humphreys, J. 'Election Manifestos: Do They Matter?' YouGov, 16 May 2011. https://yougov.co.uk/politics/articles/1684-election-manifestos-do-they-matter and Guest, Barnaby, 'Are We Voting On Loyalty Over Logic? The Real Influence Of Party Manifestos', Unlock Democracy, 20 July 2024. https://unlockdemocracy.org.uk/blog1/manifestos-voting-intention

11 Ackerman and Fishkin, op. cit., p.3

12 Smyth, Chris, and Hayward, Eleanor, 'Hospital funding will be tied to patient ratings in NHS overhaul', *The Times*, 27 June 2025

13 Social Services Inspectorate, DoH, 'Performance Ratings for Social Services in England, 2001–2'; see: https://image.guardian.co.uk/sys-files/Society/documents/2002/05/29/StarRatings.pdf

14 Cutler, T., and Waine, B., 'Advancing public accountability? The social services "star" ratings', *Public Money and Management*, 23 (2006), pp.125–128

15 Try it. It's great fun: https://www.medicare.gov/care-compare/

16 Five-Star Quality Rating System, Centers for Medicare & Medicaid Services. https://www.cms.gov/medicare/health-safety-standards/certification-compliance/five-star-quality-rating-system

17 Silver-Greenberg, Jessica, and Gebeloff, Robert, 'Maggots, Rape and Yet Five Stars: How U.S. Ratings of Nursing Homes Mislead the Public', the *New York Times*, 13 March 2021

18 'Ranked-choice voting explained: What to know for NYC mayoral election', Eyewitness News ABC7NY, YouTube, 23 June 2025. https://www.youtube.com/watch?v=6nsHLne8rrQ

19 'Lander & Mamdani Cross Endorse, Brad Lander for NYC', YouTube, 13 June 2025, https://www.youtube.com/watch?v=GYH_zerPE3M

20 Oppenheim, Oren, and Troadec, Halle, 'New York City mayoral debate: Sharp attacks as Mamdani, Cuomo, Sliwa face off a final time', ABC News, 23 October 2025. https://abcnews.go.com/Politics/new-york-city-mayoral-debate-sharp-attacks-mamdani/story?id=126772785

21 See Sunstein, Cass R., and Thaler, Richard, *Nudge: Improving Decisions about Health, Wealth, and Happiness* (Penguin, 2022)

22 'My Society: We work for a better, stronger, more inclusive democracy ... see what we've achieved!', https://www.mysociety.org

23 Mercurio, Zach, 'Why So Many People Feel Like They Don't Matter: How a "mattering deficit" is fueling widespread loneliness and despair', *Psychology Today*, 27 June 2025

24 Flett, Gordon L., 'An Introduction, Review, and Conceptual Analysis of Mattering as an Essential Construct and an Essential Way of Life', *Journal of Psychoeducational Assessment*, 40(1) (2021)

25 Arendt, Hannah, *The Origins of Totalitarianism* (Penguin Modern Classics, 2017), p.478

26 'Social connection can transform our whole health and well-being', US Dept of Health and Human Services, Reports and Publications, 2023. https://www.hhs.gov/surgeongeneral/reports-and-publications/connection/index.html

27 Ibid, citing Schram, Miranda T. et al, 'Social Network Characteristics Are Associated With Type 2 Diabetes Complications: The Maastricht Study', *Diabetes Care*, 41(8) (2018), https://diabetesjournals.org/care/article-abstract/41/8/1654/36381/Social-Network-Characteristics-Are-Associated-With?redirectedFrom=fulltext

28 Ibid, citing 'Social Support and Adherence to Treatment in Hypertensive Patients: A Meta-Analysis', *Annals of Behavioral Medicine*, 49(3) (2015), pp.307–318

Chapter Seven

1 Uber Rider Ratings breakdown example: https://www.uber.com/en-HK/newsroom/rider-ratings-breakdown

2 'Why your Uber star rating is important and how you can improve it', Inshur, 25 April 2022. https://inshur.com/en-gb/blog/why-your-uber-star-rating-is-important-and-how-you-can-improve-it

3 'How to check and improve your Uber rider rating', Uber UK, 18 June 2019. https://www.uber.com/en-GB/blog/how-to-improve-uber-rider-rating-uk/

4 If you do want to go too deep, try Axelrod, R., and Hamilton, WD., 'The Evolution of Cooperation', *Science*, 211 (1981)

5 Sapolsky, op. cit., p.345, p.393

6 'Uber driver jailed for killing passenger he ordered out of car', BBC News, 13 May 2024. https://www.bbc.co.uk/news/articles/crgyy635ezlo

7 Neath, Amelia, 'Uber sued after mother-of-four driver was shot in head by rider while pleading for her life', the *Independent*, 12 September 2023

8 'Who are Top Critics?', Rotten Tomatoes. https://www.rottentomatoes.com/critics/top_critics

9 'We just updated our Tomatometer critics criteria,' Rotten Tomatoes, 28 August 2018. https://editorial.rottentomatoes.com/article/updated-tomatometer-critics-criteria/

10 Cambra-Fierro, Jesus, Melero, Iguacel, and Sese, F. Javier, 'Managing Complaints to Improve Customer Profitability', *Journal of Retailing*, 91(1) (2015), pp.109–124

11 Olszowska, L., 'How the boss of the Beckhams' favourite restaurant is turning the tables on infuriating diners – by reviewing or banning THEM', *Daily Mail*, 12 March 2025 and Evans, H., 'Are you a 'social media scumbag'? Restaurants reviewing their customers', *The Times*, 13 March 2025

12 'Introducing GlassTable: Read Between the Courses'. https://glasstable.carrd.co

13 The literature on gamification psychology is vast, but see for example: Sailer, M., Ulrich Hense, J., Mayr, S.K., and Mandl, H., 'How gamification motivates: An experimental study of the effects of specific game design elements on psychological need satisfaction', *Computers in Human Behaviour*, 69 (2017), pp.371–380 and Chou, Yu-kai, *Actionable Gamification: Beyond Points, Badges and Leaderboards* (CreateSpace, 2015)

14 'Trustpilot: Write a Review Week'. https://business.trustpilot.com/blog/build-trusted-brand/write-a-review-week

15 Lindemann, Nigel, 'What's the average survey response rate?: Tips for consultants, marketers, HR and anyone looking to boost data collection', Pointerpro, SurveyAnyplace, August 2024. https://pointerpro.com/blog/average-survey-response-rate/

16 'How Does Vine Work?' Amazon Seller Central. https://sellercentral.amazon.co.uk/help/hub/reference/external/GX3CEV63Y4ZZAP37#mnd_2jc_jcb__section_itc_1pr_zsb

17 Gibbs, Samuel, 'Through the letterbox: the secret life of an Amazon reviewer', the *Guardian*, 28 September 2016

18 Thorvarium, 'Review Insightfulness'. https://chatgpt.com/g/g-686008d8c8a88191ae74d61410a44332-review-insightfulness

19 See r/AmazonVine on reddit.com: https://www.reddit.com/r/AmazonVine/comments/1oz4kqs/comment/npba543

Chapter Eight

1 Elassar, Alaa, 'Skin care brand Sunday Riley wrote fake Sephora reviews for almost two years, FTC says', CNN, 23 October 2019. https://edition.cnn.com/2019/10/22/us/sunday-riley-fake-reviews-trnd

2 Vahl, Shari, and Rimmer, Monica, 'Top restaurant owners blackmailed over fake reviews', BBC News, 17 July 2024. https://www.bbc.co.uk/news/articles/c51ypp5dxe5o

3 Brown, Lane, 'The Decomposition of Rotten Tomatoes', *New York* magazine, 6 September 2023

4 '22 Online Review Statistics Every Business Can Leverage in 2023', Podium, 16 March 2021. https://www.podium.com/article/online-review-statistics/

5 Dedewa et al, op. cit. 'Five-star growth: Using online ratings to design better products', McKinsey & Company Report, 12 August 2021

6 See: www.askfortransparency.com

7 Caulfield, Timothy, *The Certainty Illusion, What You Don't Know and Why It Matters* (Allen Lane, 2025)

8 'Investigating the prevalence and impact of fake reviews', UK Government Research and Analysis, 25 April 2023. https://www.gov.uk/government/publications/investigating-the-prevalence-and-impact-of-fake-reviews

9 'FTC's rule banning fake online reviews goes into effect', AP News, 22 October 2024. https://apnews.com/article/ftc-rule-fake-reviews-ban-b7e57e33c1818fa58dd742a921643239

10 'Boost Your Business: Buy Facebook Page Reviews & Enhance Your Reputation!' Digital Care Bangladesh website, live 3 March 2026. https://digitalcarebd.com/product/buy-facebook-page-reviews/

11 Shaikh, Shadma, 'Inside India's Underground Network of Fake e-commerce reviews', *Mint*, 8 April 2025

12 Muk, Isaac, 'Inside the weird, shady world of click farms', *Huck*, 22 January 2024

13 Bhandari, Bibek, 'Photographer steps inside Vietnam's shadowy "click farms"', CNN, 7 March 2024. https://edition.cnn.com/style/vietnam-farms-jack-latham-beggars-honey

14 See Lindquist, Johan, 'Illicit Economies of the Internet: Click Farming in Indonesia and Beyond', Window on Asia, *Made in China Journal*, 12 January 2019. https://madeinchinajournal.com/2019/01/12/illicit-economies-of-the-internet-click-farming-in-indonesia-and-beyond/

15 Pitrelli, Monica, 'Here's how many fake reviews Tripadvisor found on its website in 2024', CNBC, 25 May 2025. https://www.cnbc.com/2025/05/26/

heres-how-many-fake-reviews-tripadvisor-found-on-its-website-in-2024-.html

16 'Buy Five Star Verified Amazon Reviews', https://www.skitsolutionbd.com/Services/fiver-star-amazon-verified-reviews/

17 Lester, Paul, 'Fake reviews make consumers more than twice as likely to choose poor-quality products', Which?, May 2020. https://www.which.co.uk/news/article/the-real-impact-of-fake-reviews-aqg3x3G8bPpE

Chapter Nine

1 A third season launched in late 2025. https://www.bbc.co.uk/sounds/brand/m001wjgz

2 Ismail, Nick, 'The value of data: forecast to grow 10-fold by 2025', *Information Age*, April 2017. https://www.information-age.com/data-forecast-grow-10-fold-2025-5199/; also Bartley, Kevin, 'Big data statistics: How much data is there in the world?', Rivery, 28 May 2025. https://rivery.io/blog/big-data-statistics-how-much-data-is-there-in-the-world/

3 Levitin, Daniel J., *The Organized Mind: The Science of Preventing Overload, Increasing Productivity and Restoring Your Focus* (Penguin, 2015), p.377 (Kindle Edition)

4 Cited by Gleick, J., 'Surfing the data flood: Why omniscience is a curse', *New Scientist*, 30 March 2011

5 For example, 'What movie did you try the phantom menacing technique on for me it was this movie?' Reddit, 2024. https://www.reddit.com/r/Schaffrillas/comments/1cdpbuk/what_movie_did_you_try_the_phantom_menacing/

6 Janis, Irving L., *Groupthink: Psychological Studies of Policy Decisions and Fiascoes* (Houghton Mifflin, 1982)

7 Sunstein, Cass R., and Hastie, Reid., *Wiser: Getting Beyond Groupthink to Make Groups Smarter* (Harvard Business Review Press, 2015)

8 Ibid, p.209

9 Kahneman, Daniel, Sibony, Olivier, and Sunstein, Cass R., *Noise: A Flaw in Human Judgement* (Little, Brown, 2021)

10 Sunstein and Hastie, op. cit., p.102

11 For a Lillie Langtry bio try Jersey Heritage. https://www.jerseyheritage.org/history/lillie-langtry/

12 Salvi, F., Ribiero, MH., Gallotti, R., and West, R., 'On the Conversational persuasiveness of GPT-4', *Nature Human Behaviour*, 9 (2025), pp.1645–1653; or Dash et al, 'Can AI Change Your View? Evidence from a Large-Scale Online Field Experiment'. https://regmedia.co.uk/2025/04/29/supplied_can_ai_change_your_view.pdf

13 Santos., A.M., and Antonio, N., 'Improving trust in online reviews: a machine learning approach to detecting artificial intelligence-generated reviews', *Information Technology and Tourism*, 27 (2025), pp.739–766

14 Baek, T.H., Kim, H., and Kim, J., 'AI-generated recommendations: roles of language style, perceived AI human-likeness, and recommendation agent', *International Journal of Hospitality Management*, 126 (2025)

15 Dash, Saloni, Xu, Yiwei, Jalbert, Madeline, and Spiro, Emma S., 'The persuasive potential of AI-paraphrased information at scale', *PNAS Nexus*, 4(7) (2025), p.207

16 Marr, Bernard, 'Big Data: The Mega-Trend That Will Impact All Our Lives', LinkedIn post, 28 August 2013. https://www.linkedin.com/pulse/20130827231108-64875646-big-data-the-mega-trend-that-will-impact-all-our-lives/

17 Lambert, Madeleine, 'Fake AI Zillow Reviews Increased by 558% from 2019 to 2025', Originality.ai, 14 October 2025. https://originality.ai/blog/ai-zillow-reviews-study

18 'The EU Artificial Intelligence Act: Up-to-date developments and analyses of the EU AI Act'. https://artificialintelligenceact.eu

19 Ibid

Chapter Ten

1 Speech in the House of Commons, 11 November 1947

2 Coalition for Trusted Reviews. https://www.coalitionfortrustedreviews.com

Acknowledgements

Thank you to my agent Jane Turnbull for suggesting I write a book in the first place and then coming up with a winning title. Thanks to Pete Selby and James Lilford at New Modern for commissioning it and then showing me how to do it. Thanks to Nige Tassell for curtailing my ramblings, spotting cock-ups and knocking it all into shape.

The Coalition for Trusted Reviews was kind enough to let me attend their 2025 US conference where I met, spoke to and heard talks from several of the contributors to this book. Thank you to Francis Mallinson, Georgia Brown and Isabel Pyle at DGA for all of your invaluable help. I am also grateful to the Equal Vote Coalition for allowing me to use their President Pet voting scenario. Democracy is stronger for it.

Thanks to Kristen Baldwin, Ian Burdge, Michael Hogan, Andrew Pettie, Alice Rhodes, Anna Stanley, Jane Timmis and Guy Veale for suggestions, ideas and support as well as to Graham Kibble-White and Catherine Gee for your patience while I took time off to write.

Big up Shark vacuum cleaners, too.

The team at New Modern would like to thank
the following individuals:

Nige Tassell for copy-editing
Joe Barnes for proofreading
Marie Doherty for typesetting
Stuart Tolley for cover design
Lizzie Dorney-Kingdom for publicity
**Marie Lecouturier, Charlotte Rose, Andreina Brezzo
and the team at Simon & Schuster UK** for sales and distribution